Berliner geographische Studien

Herausgeber: Frithjof Voss

Schriftleitung: Christian Schulz

Band 47

Integrated Application of Geographic Information System and Remote Sensing in Solving Hydrogeological and Environmental Problems in the Central Part of Ethiopia and its Possible Extensive Future Use

von Mezemir Fikre-Mariam Wagaw

D 83 **Berlin 2002**

Institut für Geographie der Technischen Universität Berlin

Die Arbeit wurde am 22. Juli 2002 von der Fakultät Architektur-Umwelt-Gesellschaft der Technischen Universität Berlin unter dem Vorsitz von Prof. Dr. J. Küchler, Berlin, auf Grund der Gutachten von Prof. Dr. H. Kenneweg, Berlin, und Prof. Dr. F. Voss, Berlin, als Dissertation angenommen. Sie wurde gedruckt mit finanzieller Teilunterstützung der Technischen Universität Berlin. Die Abbildungen und Karten wurden auf CD-ROM geschrieben und als digitale Anlage bereitgestellt.

Herausgeber: Prof. em. Dr. F. Voss
im Institut für Geographie
der Technischen Universität Berlin
Rohrdam 20-22, D-13629 Berlin

Schriftleiter: Christian Schulz

Titelseite: 3D-view of North-East Africa - based on the e20n40 tile of the USGS

OBSERVATION POSITION:

X = 671936 m.	Y= -711611m	AGL = ASL = 92316m
FOV = 50	pitch = -30	azimuth = roll = 0
sun-azimuth = 9°	sun-elevation = 59°	sun-ambience = 0.50
DEM exaggeration = 20	view. range = 3596540m	georef.: UTM - zone 37

Inset/ Titelseite: Mezemir Fikre-Mariam Wagaw

ISSN 0341 - 8537

ISBN 3 7983 1739 9

Gedruckt auf säurefreiem alterungsbeständigem Papier

Druck: Offset-Druckerei Gerhard Weinert GmbH Postfach 420312
12603 Berlin

Vertrieb: Technische Universität Berlin
Universitätsbibliothek, Abt. Publikationen,
Strasse des 17. Juni 135, D-10623 Berlin
Tel.: (030) 314 22976, -23676
Fax: (030) 314 24741
E-Mail: publikationen@ub.tu-berlin.de

Verkauf: Gebäude FRA-B
Franklinstrasse 15 (Hof), D-10587 Berlin

To my brave twins,

Etsubdenk & Genet Fikre-Mariam.

Acknowledgements

This research work is done at the faculty Architecture-Environment-Society of the Technical University of Berlin, in Germany.

This project would have not come to its realization without the help of many people working in the Technical University of Berlin and the staff members from the Ethiopian Institute of Geological Surveys (EIGS), in the Ministry of Mines and Energy, Addis Ababa.

My great thanks goes to my adviser Prof. Dr. F. Voss. Without his help I would not be here at this time. His help in formulating the thesis research topic, as well as advises in its realization were crucial. Similarly, I would like to thank my co-advisor Prof. Dr. Kenneweg for his unreserved support, constructive ideas and comments by reading my manuscript in depth in a very short time. I admire him for the readiness he showed in taking up the advisor position after a short notice from me. My special thanks goes to Dipl. Geogr. Gabriele Fliessbach for allowing me to work in her computer lab of "Kartographie Verbund Fakultät VII der TU Berlin" and her crucial advices.

The late Prof. Dr. K.-P. Lade from the Salisbury State University, USA, was my master mentor of GIS. Before his sudden death, being a visiting professor to our faculty and my subsequent two visits to him at the Salisbury State University, he had considerably influenced the chapters 6, 7, and 8. The possible introduction of GIS to Ethiopia, seen from different implementation aspects - chapter 8, was included due to his initiation.

I am indebted to thank my colleagues from the EIGS, Addis Ababa, for their cooperation in the field work and making crucial data available for the successful completion of this research. My thanks to Mr. Ketema, Chief Geologist of the EIGS, and Mr. Gari Fufa the then head of the Geophysics department, for their support and permission of all necessary data. My greatest thanks goes to Mr. Abebe Ayele, senior expert in geophysics and the then vice head of geophysics department, for making my field work successful and for his crucial administrative and professional support available throughout my work. I would like to thank Dr. Tilahun, associate prof. of Geology and Geophysics at the Department of Geology in Addis Ababa University for his support. The enthusiasm shown by Mr. Getahun from the hydrogeology department both in commenting and aiding where possible and necessary were of great help as well. Mr. Terefe from EIGS library facilitated me a lot in the course of my literature work for which I say many tanks.

My special thanks goes to Mr. Taye Tessema from the Humboldt University of Berlin for his willingness in reading through my manuscript. I would like to thank my fellow postgraduate students for their time and effort keeping me some and helping me wherever possible especially Berhanu Ayele, Carmen Kittelberger, Pransanjit Dash, Kitaba Bersissa and Radoslav Dochev with whom I worked day to day.

Finally, I would like to thank my family whole heartedly and collectively for their unreserved encouragement, especially Birtukan Fikre-Mariam and Getachew Sebsibe for their great support in the field work during the difficult times.

Berlin – December 2002 Mezemir Fikre-Mariam

Table of Contents

Table of Figures

List of Tables

Table of Maps

List of Appendices

List of Acronyms

CCT	Computer Compatible Tape
CD-ROM	Compact Disk Read Only Memory
E, W, N, and S	East, West, North, and South (Compass direction)
ERDAS	Earth Resource Data Analysis System
ERE	Effective Resolution Element
FFT	Fast Fourier Transformation
FOV	Field of View
GIS	Geographic Information System
GUI	Graphical User Interface
IFFT	Inverse Fast Fourier Transformation
IFOV	Instantaneous Field of View
IHS	Intensity Hue Saturation
IPC	Inverse Principal Component Analysis
IR	Infra Red
N1a	Alaji basalts of early Miocene, (which forms the main trap basalt)
N2r	Old alkaline and per alkaline rhyolite domes and flows of early Pliocene
N2Qb	Bofa basalts of Pliocene
NNE	north-north-east (Compass direction)
NNW	north-north-east-west (Compass direction)
PC	Principal Component Analysis
Qwra	Alkali and per alkali rhyolites, trachytes, domes and flows of basalt of Pleistocene
RAM	Random Access Memory
RGB	Red Green Blue
TIN	Triangular Irregular Network
TCO	Total Cost of Ownership
TOP	Target (Task) Oriented Project
USGS	United States Geological Surveys
VLF	Very Low Frequency
WWS	West, West South (Compass direction)

Key words: GIS, remote sensing, information infrastructure, lineament, groundwater, environment architecture

Abstract

This research is conducted with the aim of studying the structural setup, geology and hydrogeology of the western escarpment of the main Ethiopian rift valley by using remote sensing data and GIS technology and its possible use for the dwellers of the area. A Landsat TM, MSS, SPOT panchromatic image, different topographic maps of 1:50000 scale as well as analog historical aerial photos of selected areas were used for the study. These data were processed and integrated into a single GIS database. The meteorological data from the three different stations were also processed and compiled to the database. The input and integration of all available results to a single reliable and robust database created a virtually new way for studying and analyzing a multitude of overlays and their combination.

The analysis had demonstrated that remote sensing and GIS technologies are relevant and vital instruments in mapping the main lineaments as well as better understanding of the ground water availability. Results about the village distribution, geomorphology and the continuous natural forest diminishing were also obtained. Further, it was shown that in this climate zone, the search and exploitation of groundwater should not be considered as an independent work and as a closed entity in itself. It should rather be the integral part of an overall balanced environmental management and social development of the area.

The viability of using remote sensing as a fast, timely and reliable information source was discussed. The benefit and usefulness of introducing the GIS as an interdisciplinary collective tool for tackling the diverse needs and problems was articulated in detail and further study was recommended.

Schlüsselwörter: GIS, Fernerkundung, Informationsinfrastruktur, tektonische Störungslinien, Grundwasser, Umwelt Architektur

Kurzfassung

Die vorliegende Arbeit hat die Zielsetzung, den strukturellen Aufbau und die Hydrogeologie des westlichen Steilhanges (escarpment) sowie das äthiopische Hauptgrabensystem auf der Grundlage digitaler Satellitenbilder und unter Anwendung der GIS-Technologie zu untersuchen und deren Nutzungsmöglichkeit für die Bewohner zu erkunden. Dafür wurden die Landsat TM, MSS sowie die SPOT-panchromatischen digitalen Daten, topographische Karten 1:50000 und historische analoge Luftbilder für kleinere ausgewählte Gebiete verwendet. Diese Daten wurden erfasst, digital bearbeitet, georeferenziert und in eine einheitliche GIS-Datenbank integriert. Die so erstellte einheitliche Datenbank ermöglichte unterschiedliche neue Wege der Betrachtung und Analyse in verschiedensten Kombinationsformen und Überlagerungen. Die meteorologischen Daten von drei verschiedenen Stationen wurden ausgewertet.

Diese Arbeit hat gezeigt, dass Fernerkundungs- und GIS-Technologien wichtige Mittel für die Kartierung der tektonischen Liniamente sind sowie besseres Verständnis des geologischen Aufbaues und der Verfügbarkeit des Untergrundwassers ermöglichen. Es wurden auch Ergebnisse über die Siedlungsstruktur, die Geomorphologie und den Forstbestandsschwund erzielt. Außerdem sollte in dieser Klimazone die Suche und Nutzung des Grundwasser-Vorkommens nicht als eine in sich geschlossene und unabhängige Arbeit gesehen werden, sondern vielmehr als ein integraler Teil der gesamten Umweltplanung und der ausgeglichenen Entwicklung des Gebietes.

Die Vorteile der Nutzung von Fernerkundungsdaten als eine schnelle flexible Datenquelle wurden erörtert. Die Vorzüge und die Notwendigkeit der GIS- Einführung als ein gemeinsames Informationsverarbeitungs- und Verwaltungssystem für verschiedene Fachdisziplinen und Aufgabenstellungen wurden dargestellt und darüber hinaus Lösungswege für weitere Untersuchungen vorgeschlagen.

Summary

Sufficient surface/ground water availability is one of the crucial factors for a healthy future socio-economic development in the study area. Hence, as a contribution towards a better understanding of the structural/hydro geological setup of the area, this study uses remote sensing data, and applies GIS databases and tools. Specifically, the study assesses the geologic/geomorphologic setup of the area, the alignment of fissures, faults and their formation in the region.

For this study, digital images of the Landsat TM, MSS and SPOT panchromatic were used. Additionally, analog historical aerial photographs of a smaller area as well as topographic map of 1:50000 scale were available.

For the digital image processing, the creation and analysis of the GIS database, the programs Erdas Imagine and ARC/INFO were implemented. The main research aims were to study:

- the construct and composition of the upper (shallow) geology,
- the tectonical structure and its distribution pattern in the region,
- the underground water circulation,
- the dwelling pattern, geomorphology, surface water, and their interaction,
- the discovery of the above geology and its potential utilization by the villages in the respective adjacent areas as well as
- commenting the suitability of remote sensing and GIS for similar problem settings.

A detailed study on the lineament pattern was carried out. In this study the tectonic lineament of the NE-SW was found to be the predominant one, followed by the semi perpendicular NW-SE direction. Next to that, the population distribution in the area was studied. Here, it was easily revealed that the great majority of the villages are settled on the tops and sides of the volcanic cones and the top of the horst formations. The distribution of the villages in the study area is seen to be uniform with a higher concentration around the towns and irrespective of the climate regime.

After that rivers, wadies and wetland areas were mapped. With the help of the GIS technology various analyses were carried out, and the interactions among these quantities were commented and discussed. Further, the village distribution pattern, the water availability and the natural vegetation were brought in relation to the slope and aspect of the area. Streams, rivers, wetlands, and the forest distribution were overlaid and studied. Their interdependency was analyzed, interpreted and discussed.

The forest and village distribution maps were overlaid and compiled. The steadily diminishing size of the forest, which is mainly caused by the high demand of fire and construction wood is presented and articulated. Unfortunately, against this alarming devastation of the native forest, there is still no meaningful re-foresting program in place.

For parts of the study area, the digital satellite image processing is shown to be a vital supplementary information source to those already existing geological/ hydro geological maps. This has resulted in a considerable information gain.

From the study, the following conclusion were derived:

- especially in the escarpment area, but also in the rift valley region to a lesser extent, there are substantial parallel tectonic lineaments which are mainly northeast - southwest directed,
- there are also second group of lineaments, with lesser intensity, in the NW-SE direction,
- the lineament length varies from few kilometers to 50 km or more, and
- it is also observed that the overwhelming majority of the villages are located on the tops and sides of horst formations and volcanic eruption centers.

The existence of such tectonic structure may create a favorable condition for ground water circulation in the region. The lineament structures and the weak zones can potentially be used as an underground reservoir. The effective use of such structures may increase water quality and decrease the loss of water in form of evaporation. In this regard future additional high resolution study is vital.

The locations with high rainfall intensity, mainly the escarpment areas, are at the same time areas of extensive agriculture, causing a concurring interest for the same locality with the new-water building and surface water catchments. In order to avoid any negative environmental impact, the extensive agricultural use should be brought in harmony with the water protection and environmental preservation locations.

Based on the results, an integrated approach to the surface/groundwater as a single entity is recommended. The approach of building “many small” dams more frequently upstream at the valleys - on the escarpments is more preferable. This causes the deceleration of the erosion, and less evaporation. The availability of water for villages around these high land localities will be secured, the villages in the lowland area and in the rift valley can get clean water in a form of ground water in their vicinity. In such a method, the risk of mineralization on the rift floor could also be confined. Besides this, a wide possibility may be opened for the construction of several “decentral and small” hydroelectric power-stations for local consumption.

An additional aim of the study was to see how effective the used remote sensing and GIS technologies for hydro geological and environmental problem settings could be. Here the remote sensing is proved to be a fast and flexible data source. The GIS is also found to be an effective geographic (including the respective trailer attribute) data management and analysis tool.

Some methodical approaches were discussed on how to tackle the effects of cumulative impact on the environment in Ethiopia. In this respect the collective approach of GIS-introduction and its consecutive collective implementation in Ethiopia was seen in depth. The proper introduction and utilization of GIS and related technologies can serve as an information infrastructure tool, and this may decisively contribute to the development of the country and can be a great help for the government’s declared “sustainable development and poverty reduction” program which is underway in the recent years. Towards this end, an integrated multi-user introduction of GIS platform for the different specialists and users was discussed and recommended.

Zusammenfassung

Eine ausreichende Qualität und Menge an Oberflächen- bzw. Grundwasser ist einer der Hauptbausteine für eine anhaltende und gesunde gesellschaftliche Entwicklung des Untersuchungsgebietes. Deshalb setzt diese Arbeit die Geofernerkundung und das GIS-Verfahren ein, um einen Beitrag zum besseren Verständnis der strukturellen und hydrogeologischen Gegebenheiten und Gesetzmäßigkeiten des Untersuchungsgebietes zu leisten.

Die digitalen Bilddaten der Landsat TM und MSS sowie der panchromatischen SPOT-Aufnahmen wurden dazu verwendet. Als Referenzbasis diente die topographische Karte 1:50000. Für eine gezielte Spezialuntersuchung wurden historische analoge Luftbilder herangezogen. Für die digitale Bildbearbeitung, die Erstellung der GIS Datenbank sowie der späteren Analyse wurden die Programme Erdas Imagine und ARC/INFO verwendet. Im Mittelpunkt der Untersuchung standen:

- der Aufbau und die Zusammensetzung des Untergrundes,
- die tektonische Struktur und deren Verteilungsmuster in der Region,
- die Untergrundwasserzirkulation, deren Erschließung und Nutzungsmöglichkeit für die angrenzenden Dörfer,
- die Siedlungsstruktur, Geomorphologie, Oberflächengewässer und deren Wechselwirkung sowie
- die Tauglichkeit von Fernerkundung und GIS für solche umweltbezogene Aufgabestellungen.

Zunächst wurden die NO-SW gerichteten tektonischen Lineamente, die im Untersuchungsgebiet am häufigsten auftreten, sowie die weit weniger vorkommenden NW-SE gerichteten Lineamente eingehend analysiert. In einem zweiten Schritt wurde die Bevölkerungsverteilung im Untersuchungsgebiet analysiert. Dabei konnte festgestellt werden, dass der größere Teil der Dörfer auf den Vulkankegeln und entlang der Horstbildungen in den Steilhangs-Gebieten angesiedelt ist. Die Dorfverteilung im Untersuchungsgebiet ist, mit etwas größerer Dichte in der Nähe der Kleinstadtsiedlungen und ungeachtet der klimatischen Zone, weitestgehend einheitlich.

Danach erfolgte die Kartierung der Flüsse, Wadis, und anderer Feuchtgebiete. Mit Hilfe von GIS Methoden wurden umfangreiche Analysen vorgenommen und die jeweiligen Wechselwirkungen aufgezeigt und erörtert. Die Größen wie Hangrichtung (aspect) und Hangneigungswinkel (slope) wurden aus den digitalisierten Höhenlinien erstellt und in Verbindung mit der Verteilung der Dörfer, der Vegetationsstruktur und dem Gewässernetz diskutiert.

Wald- und Dorfverteilungskarten wurden überlagert und verglichen. Es wurde auch der Forstbestandsschwund - der überwiegend durch den sehr hohen Brenn- und Bauholzbedarf verursacht ist - dargestellt und problematisiert, wogegen es bis jetzt keine nennenswerte Wiederaufforstungsaktivität gibt.

Für einen Teil des Untersuchungsgebietes wurde aufgezeigt, dass der Satellitenbild-Auswertung ergänzend zu den vorhandenen geologischen und hydrogeologischen Karten eine erhebliche Bedeutung für die Informationsgewinnung zukommt.

Aus dieser Arbeit können folgende Schlussfolgerungen gezogen werden:

- besonders in den Steilhangbereichen - in dem Grabensystem in geringerer Intensität - sind die NO-SW gerichteten tektonische Lineamente gut ausgebildet und ausgeprägt,
- es ist auch eine zweite Lineamentenrichtung, die NW-SE gerichtet ist, festzustellen,
- die Länge dieser Lineamente reicht von einigen Kilometern bis 50 km oder mehr, und
- die meisten Dörfer sind auf den Höhen und Seiten der Horstformationen und auf den Hängen von den jetzt passiven Vulkanausbruchszentren lokalisiert.

Die Existenz solcher tektonischer Strukturen kann eine sehr gute Bedingung für die Grundwasserleitung und -speicherung schaffen. Diese Strukturen können als Wasserspeicher gezielt genutzt werden. Damit könnten hohe Wasserqualität und geringere Verdunstungsverluste erreicht werden. Detailuntersuchungen in dieser Richtung sind angebracht.

Das Steilhanggebiet ist das Gebiet mit ergiebigen Niederschlägen, aber zugleich ein Gebiet der extensiven Landwirtschaft. Um die wahrscheinlichen Konflikte der Landnutzung-interessen zu vermeiden, sollte die landwirtschaftliche Tätigkeit im Einklang mit der Schutzgebietsbildung gebracht werden.

Ausgehend von den oben geschilderten Tatsachen ist eine Gesamtbetrachtung der Oberflächen- und Untergrundgewässernutzung zu empfehlen. Die Möglichkeit der Bildung von „vielen" Ministaudämmen in den oberen Talbereichen der Steilhänge ist im allgemeinen zu bevorzugen. Dieses wird die Erosion der landwirtschaftlichen Nutzflächen verlangsamen, den möglichen Verlust durch Evaporation verringern, die Verfügbarkeit des Wassers für die Einwohner der höheren Lagen sichern und durch die natürlichen Grundwasserzirkulations-prozesse ein sauberes Grundwasser für die Dörfer der tieferen Gebiete sichern. So eine Gesamtlösung würde zudem die Gefahr der Mineralisation der Dämme in den tieferen Gebieten weitestgehend minimieren. Außerdem ist die Möglichkeit für den Aufbau von „dezentralen kleinen" Wasserkraftwerken, die den lokalen Energiebedarf decken können, dadurch gegeben.

Ein weiteres Ziel der Arbeit galt der Prüfung der verwendeten Fernerkundungs- und GIS-Verfahren für die hydrogeologischen und umweltbezogenen Fragestellungen. Dabei konnte die Eignung von Fernerkundungsmethoden als schnelle und flexible Datenquelle nachgewiesen werden. Die Tauglichkeit von Geofernerkundungs- und GIS-Verfahren für hydrogeologische und umwelt bezogene Problemstellungen wurde eingehend diskutiert.

Es wurden verschiedene methodische Ansätze vorgeschlagen, wie die negativen kumulativen Wirkungen - die in dieser Arbeit thematisiert wurden - aufgefangen und schrittweise gelöst werden können. Die erfolgreiche Einführung und Anwendung von GIS und in deren Zusammenhang stehenden Technologien können einen entscheidenden Beitrag für die Entwicklung des Landes leisten und die effektive Bildung einer Informationsinfrastruktur ermöglichen. Für das, von der äthiopischen Regierung seit kurzem ins Leben gerufene „poverty reduction" Programm kann so ein System wichtige Grundlage bilden. In dieser Hinsicht wurde die integrierte Mehrnutzereinführung von GIS-Plattformen für Interessenten aus verschiedenen Bereichen erörtert und vorgeschlagen.

1 Problem Overview and Objectives of the Research

Ethiopia with its peculiar highly undulating and tilted geomorphology poses a serious challenge in securing enough water to its people and in executing medium to large scale economic developmental undertakings. In the last several decades, it is observed that different region of the country is struck by "major cyclic" drought repeatedly. This cyclic recurrence of "irregular rainfall" necessitates, to search for and develop a long-term solution which should address the potential scarcity of surface and ground water.

It is a well established fact that more than 85% of the Ethiopian population live in the countryside and is predominantly engaged in mixed cattle breeding and crop farming, which directly depends on the natural rain distribution and intensity. As a drinking water source, in the most traditional way, people use springs. Where this is not available, hand-dug wells and rivers are the main sources. In the rural areas, especially in the highland areas, the increased use of highly steep slope locations for plowing and as a grazing field is widely observed, which otherwise would have served as a new water building zone. This has led to a rapid deforestation and higher run-off of the surface water to the lower elevation. There is as a result often less free space, less wetland and drinking water entry area and a very limited amount of natural forest. Natural forest with its bio-diversity, except the few commercial eucalyptus tree patches, is becoming steadily scarce. The irregularity of the rainfall, increased necessity of water security for drinking, cattle, and other agricultural activities - together with the substantial population growth - require water conservation, protection and management policy at the federal and local governmental levels. The recurrent drought in all climate zones of Ethiopia forces the eventual creation of a mechanism for protection of catchments, main streams as well as locations with a very high inclination slope.

In general, the "normal" climate nature of the study area shows a prolonged dry season followed by an intense rain over a two to four month period.

The water balance problem of the area can be stated as:

i. intense rain occurs mainly in the highland areas, and to a lesser extent in lowland areas for short periods, separated in both cases by a long dry spell. The relatively short and intense storm and flood periods are within a couple of weeks over and much water tends to pass quickly out of the area,

ii. heavy concentration of runoff causes major erosion and soil degradation problems in the highland areas and at the same time the sediment transported, often with a very high speed downwards to the lowland areas, will cause major damage of cultivated land and siltation of dams,

iii. the continuing population growth and the economic activities in the rural and urban areas has considerably increased the need of water for drinking, agricultural and industrial purposes, and

iv. the periodic rain time "irregularity" in almost all parts of the country, and the tremendous social and economic disorder for millions of citizens as a direct consequence of this.

The aforementioned water cycle and the potential water shortage are crucial issues for the entire society, independent of their social structure. This problem should be encountered with an accelerated strategic approach to tackle the water, energy and environment management as a single entity. For this the proper understanding of the environment architecture – which should comprise the geology/geomorphology, native natural flora and fauna, climate regime and meteorology of a given local region - as a single entity is decisive. Effective solution of this problem demands an interdisciplinary approach from several sides of natural and social science disciplines. It requires also a substantial amount of data and information. The proper and effective management of environmental data, information and their proper access by the individual field specialists, interested governmental and non-governmental institutions, and individual end users will - in the future - determine the wellbeing and progress of the society in this climatic zone.

Often at different levels of government - decisions on capital investment and resource development for water and environmental management at large - are made in an atmosphere of greater or lesser uncertainty and with value judgments strongly conditioned by weaker information base. As a result, several development projects will have the fate of unexpected and often undesirable outcomes. As a result they are economically, socially, and environmentally unacceptable. The ever increasing demands on the natural resource needs wise and prudent management of the natural resources and the environment. Such management needs are best served if accurate, on time, and consistent resource inventories are made available to the resource manager - and any decision maker for that matter - at suitably frequent intervals, and with regular updates.

Although the identification, measurement, and inventory of our environment resource is a complex task, the technology of remote sensing, digital image processing and Geographic Information System (GIS) does offer the potential to produce a broadly consistent data base at a spatial, spectral, and temporal resolution, which is useful for a competent management. The coupling of remote sensing with GIS may basically change our methods and models of data analysis, as well as our perception of the environment and the society. The GIS allows us a very high data integrity, actualization capability, and high-grade data management and analysis facility. It allows us an "unlimited" scalability and data integrity. This leads each development task to a well coordinated Target (Task) Oriented Project (TOP), from planning, project execution and final documentation to a far later operational control of any realized and completed environmental/social project. It can permit good performance in planning, execution, documentation and the fostering of various environmental resource works in general and water supply, construction and maintenance in particular. It may also secure a transparent background information for post-construction operational management and optimal use of erected structures, which will be a distinct advantage of the GIS technology against the recent practices.

For this study remote sensing and other ancillary data were made available. Digital image processing and GIS technologies have been used. As a final result, a GIS database with more than 5 GB data size was built and maps are generated and discussed.

1.1 Introduction

The study area is located at 8°30'N and 9°N latitude and 38°40'E and 39°30'E longitude. It extends from the highlands of the escarpment in the north to the main rift valley in the south. The climate is humid, in the northern escarpment area to - semiarid, in the south - the rift valley. It is densely populated mainly by private-subsistence farmers which use mixed crop and cattle breading as their main economic means.

The humid/semiarid climate of the study area, the extensive and unplanned nature of the subsistence agriculture - causing the conversion of wetland and other environmentally sensitive locations into farm/grazing land - as well as the substantial loss of the natural forest (which otherwise played a key role for the fertility and permitted a healthy agriculture), have resulted in scarcity of springs and water bearing wadies, high rate of erosion and rapid degradation of agricultural fields. As a result, the sustainability of social development and availability of natural resource is endangered and an environmental deterioration surfaced.

The depletion of the natural forest area and the increased integration of wetland and other environmentally sensitive locations into active agricultural area have posed a difficult problem in sustainability of the rural social development.

The recurrent rain "misplacement" and drought - in wide areas of the country - have increased the demand for information on the availability of ground water and possibility of its exploitation. Additionally, the increasing demand for a variety of agricultural products and the wish to use "rain independent" irrigation agriculture has increased. A sound solution for such problems could not be effectively materialized due to the absence of crucial information about the distribution and formation of the under-ground structure and how effectively each of the villagers may use their farm plot and surface/ground water resource in its integral form.

In recognition of the importance of the geological/hydro geological investigation and mapping for drink water security and for other purposes, in the last several decades the Ethiopian institute of geological surveys had done a country wide small scale mapping and produced a hydro geological and geological map of Ethiopia with a 1:2 million scale. Detailed mapping and investigations were also done in some locations, including part of the study area Nazereth, at a scale of 1: 250000. However, the major part of the country does not have such a relatively large scale thematic map. The absence of such vital information creates a great problem in the proper management and balanced use of natural resources, especially ground/surface water. Under such uncertain condition planning, realization and reinforcement of projects would not be feasible.

The integrated application of GIS and remote sensing, by using different image data from different sensing satellites, may serve as a solution for filling this information gap. The recent accelerated development in the satellite technology as well as in the software and hardware sector may allow the delivery of even more reliable and timely results as required. This study, therefore, aims at providing information on the geology and the structural framework of the study area with a particular emphasis on the possible relations between locations and alignments of volcanic centers, the directions of the major tectonic features and the geometry of the rift's margins to the ground water circulation. Extensive study on the dwelling pattern of the villages and on the meteorology of the area was done.

For this purpose, different flowchart-models for executing practical digital processing and interpretation works were developed and used. Based on them, diverse analysis was performed.

First the regional structure was studied using the whole scene of the Landsat MSS and TM bands. Different digital processing including unsupervised classification, band rationing, PC, IPC, FFT, IFFT and IHS were applied. By convolving with 101x101 moving average convolution matrix and its subsequent processing using IHS and band ratioing brought better result in the structural and vegetation study. It was found that the upper (northern) part of the study area is more dominated with linear structures than the rift valley area. These structures are best shown on the escarpment using the 101x101-convoluting matrix, whereas in the rift valley 31x31 moving average matrix delivers better result. In some parts of the Landsat image, the hydrographic network is enhanced by thin forest galleries, dense grass cover and by shadow effects. Its orientation and density is strongly influenced by the fracturing pattern in alignment with the main stress-axis of the rift valley.

Ancillary information from the digitized topographic maps of 1: 50000 were vectorized. The vectorized coverage was further processed. TIN and different GRID coverage's were built. The information from the vectorized topographic map and the result from the image processing were integrated into the newly created GIS database.

Based on these, a number of recommendations for the surface/ground water management were discussed and further higher resolution study recommended.

Extensive discussion over the integrated and more collective use of GIS and remote sensing among different disciplines - with the aim of cost minimization and better efficiency - was done and recommendation forwarded.

1.2 Organization of the Thesis

This thesis is divided into 10 chapters including this introduction. The second and third chapter discusses location of the study area, available data, issues of earlier investigation, literature review about the geology and an investigation on the climate of the study area.

Next in chapter 4, a condensed literature review relevant to the image acquisition, processing and interpretation, together with a computer aided data manipulation procedures were briefly discussed. The results of the image processing with the help of Erdas Imagine, was presented in chapter 5.

Chapter six presents a short introduction to GIS, and the developed flow-diagrams for the practical processing. In the next chapter - chapter 7- by using the ARC/INFO software, a GIS database was built, results interpreted and thematic maps of different scales produced.

In chapter 8 a literature study and general discussion on the contemporary development challenge of the country including the possibility of a collective use of GIS in Ethiopia - under the existing financial and trained manpower constraint - was done.

Chapter 9 provides a detailed discussion on the study findings. The last chapter discusses in detail the conclusions from this study and gives some recommendations for further future works.

2 Location and Earlier Investigation

2.1 Location of the Study Area

The study area is located in the central part of Ethiopia. This study was performed, in the first part, on the full Landsat MSS and TM scene as a small-scale regional geological and geomorphologic information source. Based on this and in order to use additional data in the analysis, a smaller area of 6 topographic-sheet cutout was taken. The cutout for the detailed study is located between 8°30'N and 9°N latitude and 38°40'E and 39°30'E longitude, that lies on the western escarpment of the main Ethiopian rift valley, [Figure 1].

This area is one of the most dynamic areas of the country in terms of economical and political activities. The capital city of Ethiopia, Addis Ababa, lies just north west of the study area.

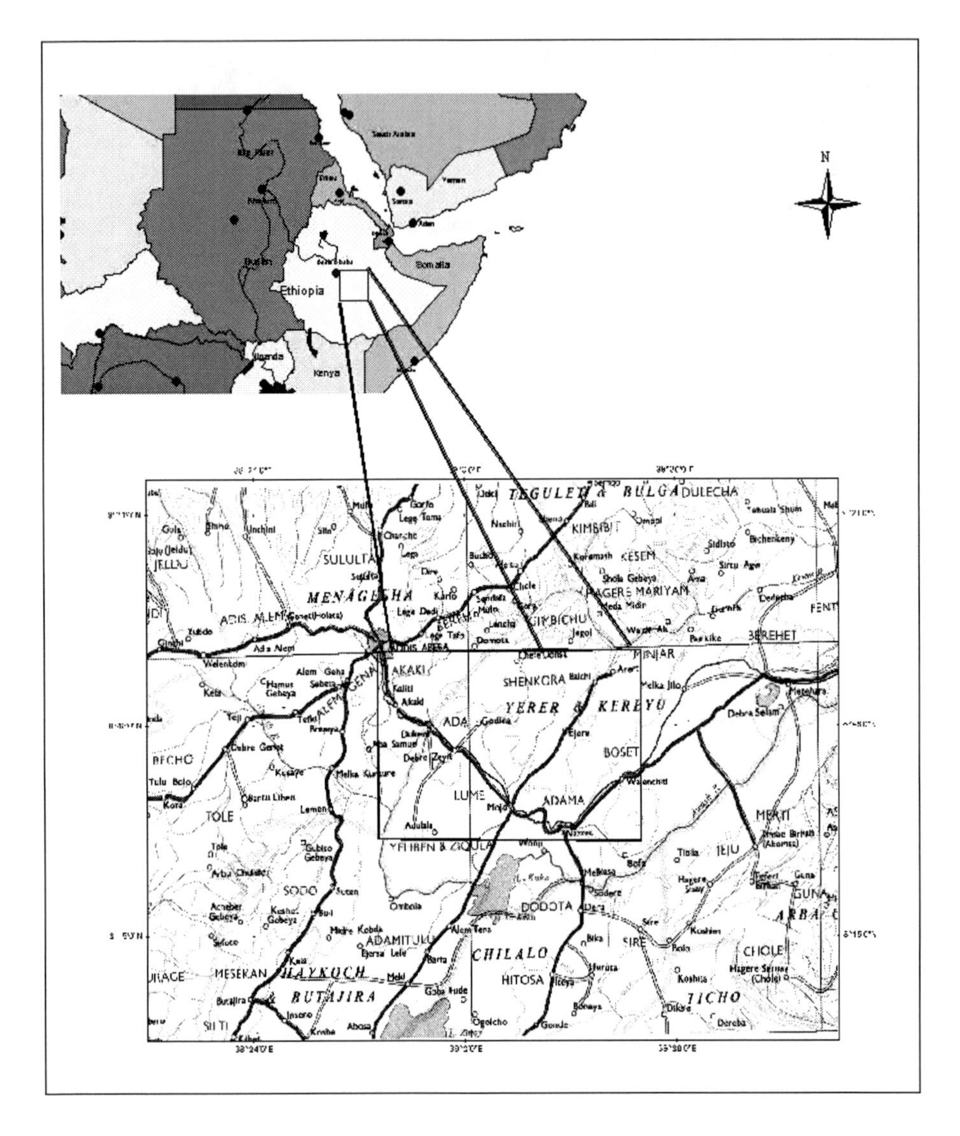

Figure 1. Study area location in the central part of Ethiopia, based on the ESRI Inc. Arcview vers. 3.1. database from 1998 and the topographic map of SE Addis Ababa, 1994, from the Ethiopian Mapping Authority with the original 1:1 million. scale.

2.2 Available Data for the Study

For this study, satellite images, aerial photographs and meteorological data were available. The original full scene Landsat TM and MSS as well as SPOT were purchased privately from the Daimler-Benz Aerospace, Dornier Satellitensysteme GmbH, 88039 Friedrichshafen. It covers to the north the watershed of the Blue Nile and to the south-south east the rift valley drainage system. This area lies approximately between 8° and 10° latitude and 38° and 40° longitude.

i) The available satellite image data includes:

- a digital Landsat MSS full scene 168/54 dated 21.04.1984 on a CCT,
- a digital Landsat TM full scene 168/054 dated 5. Jan. 1986 on CCT, and
- a digital SPOT panchromatic scene 138-332 dated 17.06.1991 on CCT.

ii) The following historical aerial photographs and topographic maps were made available by the Ethiopian Institute of Geological Surveys in cooperation with the Ethiopian Mapping Authority:

- aerial photographs of a selected area,
- topographic maps of 1:50 000 scale,
  - 0838 B1, 0838 B2 Addis Ababa,
  - 0839 A1 Chefe Donsa,
  - 0838 B4 Zikuala,
  - 0839 A3, 0839 A4 DebreZeit,
  - 0839 C2 Balchi,
- topographic map 1:250 000 central Ethiopia region,
- topographic map 1:1 million central Ethiopia- Addis Ababa,
- geological map of Ethiopia,
  - 1:2 million scale new edition 1996,
  - 1:2 million scale old edition 1981,
- hydro geological map of Ethiopia,
  - 1:2 million scale,
- geological map of Nazereth area 1:250 000 scale and
- hydro geological map of Nazereth area 1:250 000 scale.

The meteorological data for the stations Addis Ababa, Nazereth and Debre Zeit are from the Ethiopian meteorological agency in Addis Ababa - collected for approximately 30 years. The DEM, tile E020N40, was downloaded from the US Geological Survey home page, http://edcdaac.usgs.gov/gtopo30/e020n40.html .

2.3 Earlier Hydro geological Studies in the Area

There are some extensive studies performed in different parts of the study area. [Vernier1985] performed a hydro geological investigation in the Addis Ababa area. From their study they concluded "the potential occurrence of groundwater reservoirs seems to be sufficient to give a reasonable alternative water source to industrial factories". They also stated that the problem of environmental pollution might undermine groundwater exploitation. [Tsehayu1990] further discussed the potential pollution danger and the necessity of capacity building in this respect which clearly shows the need for a systematic and consistent approach to waste management and environment protection.

[Hadwen1975] gave a list of the available wells throughout Ethiopia with extensive bibliography. [Eshete1982] made a water supply study for the Debre Zeit flour miles and macaroni factory and gave a brief description of that locality. He placed the depth of the ground water between 2 and 120 meters below the surface and identified four aquifer types ranging from very good to poor potentials. He labeled the scoraceous basalt and other faulted and fractured basalt in and around Debre Zeit as the best aquifers of the location.

[Solomon1974] gave a brief discussion of the hydro geological and meteorological conditions around the Debre Zeit area. [Caldini1987] did a reconnaissance study on the wider rift valley and the Danakil depression area and gave an extensive discussion.

Different large-scale investigations for ground water and engineering geology were performed at different localities in the study area. The unpublished works of [Melaku1982] and [Paola1970] gave a description on the different geological setup of the shallow underground in different main rift valley localities.

With the help of geophysical methods, specifically dipole-dipole profiling and vertical electrical sounding, [Bohumir1982] performed hydro geological investigation in the Nazereth area. In their study, it was stated that:

- many covered depressions in the bed rock and faults were located which are considered suitable locations for groundwater development,
- the thickness and lateral variations of the lacustrine sediments in the north basin are 32 meters and a volcanic rock with an average thickness of 231 meters underlie the lacustrine sediments, and
- the thermal water table is considered to be immediately below the volcanic rocks.

As to the climate zone classification, [Melaku1982] gave three distinct climate zones, namely:

- those places above 2500 meters a.s.l., which are cold and wet highland areas, known as "Dega",
- the areas between 2500-1800 meters a.s.l. have tropical to sub-tropical moderate temperature which are called "Woina Dega", and
- the third climatic type is the "Kolla" climate, which is found in humid to arid climate zone with elevations between 1800-1500 meters a.s.l. which covers an extensive part of the rift floor area.

The Amharic terms "Dega", "Woina Dega," and "Kolla" are the equivalents of "cool temperate", "temperate" and "warm" climate zone respectively and are used officially by the Ethiopian national meteorological agency.

The above author showed that evaporation rate is, on average, less than that of the precipitation on the highland areas while it by far exceeds precipitation in the rift floor areas as shown in the [*Table 1*] below.

*Table 1. Evaporation and rainfall at Addis Ababa, Holota and Koka stations for the years 1970 to 1975 after [*Melaku1982*].*

		years					
meteorological station	pan evaporation/ rainfall (in mm)	1970	1971	1972	1973	1974	1975
Addis Ababa		1157	126.9	697	-	-	-
Holota		459.5	491.7	435.7	552.7	-	-
Koka dam	evaporation	1212	1981	2267	2107	2121	1834
Addis Ababa		1270	995	937	-	1119	918
Holota		1092	2306	2452	2012	1059	1075
Koka dam	rainfall	631	758	20.7	254	415	294

In his study of the rift valley region in Tanzania, [Mbiliny2000] has discussed the land use and the extent of land degradation by implementing the remote sensing and GIS techniques. He forwarded in his conclusion valuable recommendations on water management which may find application in the whole rift valley system, including this study area.

2.4 Development of Mapping and GIS in Ethiopia

The application of an aerial photograph mainly panchromatic has a long tradition of use in Ethiopia. In 1957, almost the entire country was covered by 1:50 000 photography in conjunction with 1:250 000 topographic mapping project. Since the beginning of 1980, the Ethiopian mapping authority coordinates the planning, creates and distributes maps and other surveying results, as shown in the [EthiopianMappingAuthority1990]. In its new 1995 revisions of the 1:50 000 topographic maps, the satellite image data was also used.

There is 1:250000 topographic map coverage for the entire country since 1972. A larger scale coverage at 1:50 000 is also available for part of the country. In all these processes, the monochrome aerial photography has played a considerable role.

As can be seen from the home page of the Ethiopian Mapping Authority, there is a short-term plan to reinforce the digital cartography and GIS for the whole country, http://www.telecom.net.et/~ema/ema.htm.

The ministry of mines and energy has produced a geological and Hydro geological map of Ethiopia at a scale of 1:2 million and few other at a scale of 1:250 000. In the past decade, the Ethiopian Institute of Geological Surveys (EIGS) had performed several feasibility and regional mapping activities in various parts of the country. These studies include all geo-science disciplines and the photo geology.

Hitherto, there is no consolidated GIS application in the country. There are now an increasing number of different research works with integrated application of remote sensing and GIS in Ethiopia from different application aspects, see [Berhe1991], [Meissner2002], [Imkemeyer2000]. Even though it was not directly on Ethiopia, the interdisciplinary geosciences research under "Sonderforschungsbereich 69" of the TU-Berlin, FU-Berlin and TFH-Berlin with the title "Geowissenschaftliche Probleme in ariden und semiariden Gebieten" had brought a comprehensive information and understanding about the northeast Africa geology and hydrogeology, see [Kenea1997], [Koch1996], [List1993_3], and [Schandelmeier1990].

2.5 Rural Agricultural Activity and the Village Distribution

There is a considerable diversity in agricultural potential and the nature of agriculture across the study area. The agriculture is dominated by small-scale mixed crop and livestock subsistence farms, usually operating on less than 1 hectare land. The most important food crops are teff (Eragrostis abyssinica), wheat, barley, maize, bean, peas, sorghum and various oilseed plants. These food crops are at the same time the dominant cash crops. It is a long-established culture for each villager here to have a garden where vegetables such as cabbage, potato, tomato, onion, and garlic are planted mainly for family but also market consumption.

The other agricultural activity sector is cattle breading. Livestock, especially cattle and small ruminants are very important for the peoples income. Each farmer family is expected to have oxen (the primary means for plough and bringing in the harvest), cows, and sheep. In warmer zone there are more goats. In some households there may be horse, mule and donkey used as the main transportation means for man and goods.

In general terms, development pathways for the farmers of the study area represent common patterns of change in economic livelihood strategies, such as continued semi-subsistence mixed crop/livestock production and commercialization of high-value perishable crops for the nearby towns and cities. These development pathways will be largely determined by three main factors which may give comparative advantage among the villages, namely:

the agricultural potential (including water security),

access to market, modern infrastructures and

population density.

The scattered village pattern and the concentration of villages on the top of ridges and hills pose a difficult task in securing clean water and enforcement of appropriate environmental management mechanism, as will be discussed in detail in the last chapters.

3 The Geology and Climate of the Study Area

3.1 Literature Assessment on the Geology and Geomorphology

In the past decades, there has been an increasing number of publication about the geology of Ethiopia and the main Ethiopian rift valley in particular. Hereafter, a condensed literature review and summary about the geology of Ethiopia and the study area is presented. For additional information, references are made to the respective literatures.

3.1.1 The Geology of Ethiopia Prior to Miocene Rifting

The Paleozoic, extending from the Cambrian to the Permian period, was an era of non-deposition in Ethiopia. The Arab-Ethiopian block was largely subjected to denudation throughout the Paleozoic. The three Mesozoic marine formations known in Ethiopia are:

- the transgressive sandstone which was deposited upon the Precambrian rock with great unconformity,
- the Antalo limestone, and
- the regressive Jurassic/Cretaceous to upper sandstone, each with temporal variations along the northwestern axis of transgression and regression.

Following the late Mesozoic early Tertiary regression of the sea to the southeast towards Somalia, in the upper Eocene and perhaps extending to early Oligocene the immense uplift of the Arab-Ethiopian swell had occurred. The sub-horizontal nature of the uplifted strata indicates the movement of an epirogenetic character. The greatest intensity of uplift in the horn of Africa has occurred in the Ethiopia's main plateau. Nearly all major rivers in Ethiopia flow down the dip slope or else in basins formed as minor down warps in the uplifted surface, [Pilger1976].

This uplift was accompanied with volcanic eruptions, which resulted in the outpouring of immense flood basalt, widely known as trap series. The trap series consist of a very thick series of lava flows, with trachytes and rhyolites occurring often near the top. The thickness of the basalt varies between few centimeters to several tens of meters, [Figure 2]. The trap series were later cut by the rift system faulting in the Miocene. Following the upper Eocene uplift and the late Oligocene to early Miocene eruptions of the trap series, mainly of Alaji basalt material, came the formation of a complex pattern of major fractures which created down faulted rifts in the late Miocene. These rifting movements may be related - though not coinciding - with the axes of maximum uplift, and it separated the once continuous Ethiopian plateau into the now existing east and west escarpment of the main Ethiopian rift valley, [Mohr1971].

a)

b)

Figure 2. Outcrops of flood basalt of the trap series NE of the TM scene a) in the Debre-Libanos area and b) in the Shenkora river valley some 100 km east of Addis Ababa.

Geo-Tectonic in the Tertiary and Quaternary

The rift system is formed by a complex pattern of narrow belts of parallel faulting. This consists of sunken stripes of land between two faults giving the characteristic rift valleys. Each pair of faults have an opposing downthrows where parallel faults oppose each other with an up throw as a raised block or as a horst.

The general geomorphologic form of the rift system and its surrounding escarpments are shown in the draped DEM model, in [Figure 3] below. In this figure, the low land and potential sedimentation areas are represented in darker while the highland areas, mountaintops and erosion-endangered locations are represented with a brighter tone. The rift valley is sunken with respect to the eastern and western escarpments and is a relatively flat area, which incorporates several younger volcanic cones. The eastern escarpment is more clearly delineated than the western escarpment.

Owing to the varied morphology of the area under consideration, many types of "local" deposits had been formed during the Miocene to recent time. In the rift system especially, the sequence is complicated by the irregular presence of volcanic rocks, including lava of varied petrography. Renewed rifting movements with associated volcanism began in the Pleistocene and have contributed to the present-day morphology, [Mohr1971].

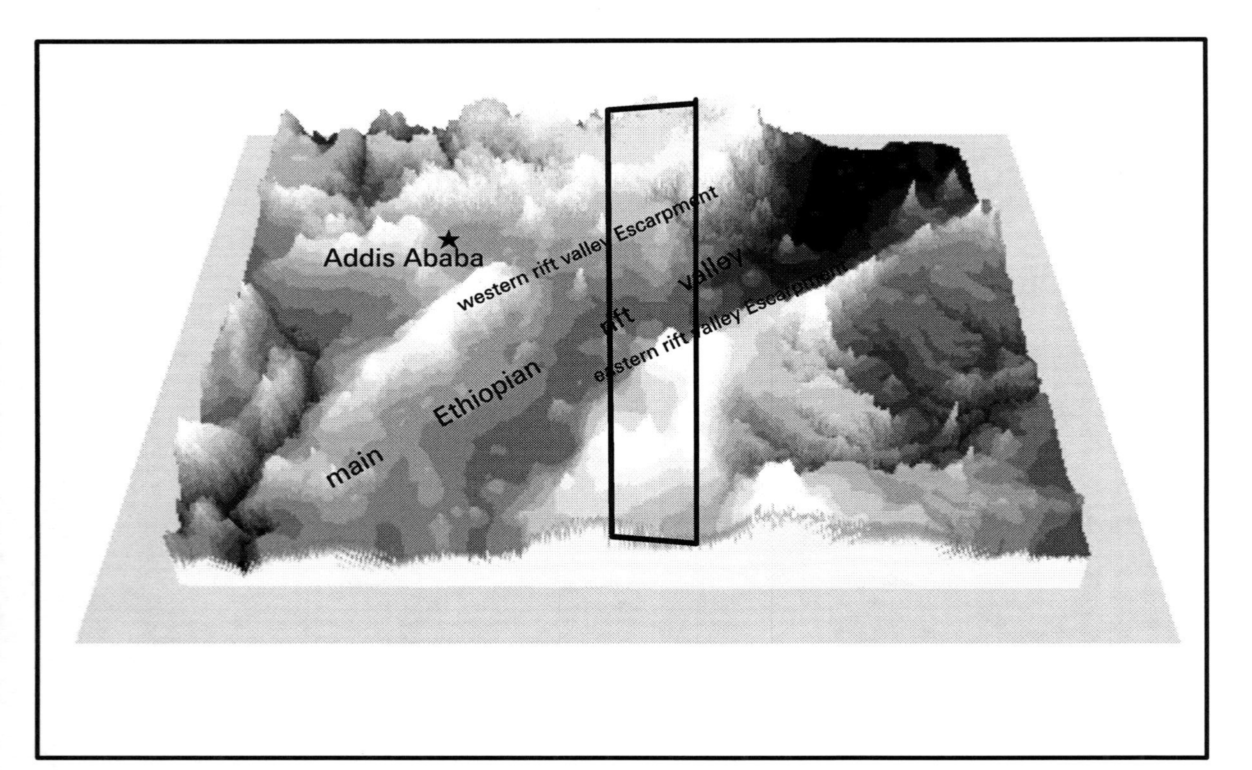

Figure 3. Observation from the south on the TM full scene area, using the 1:250000 USGS DEM data.

It shows the physiography of the central part of Ethiopia.

OBSERVATION POSITION:

...x= 509437 meters, y= 420090 meters

...AGL= 26293 meters, ASL = 26293 meters

..Direction:

...FOV: 50, Pitch = -45

...Azimuth: 0, Roll = 0.

SCENE PROPERTIES:

...DEM Exaggeration: 20

...Viewing Range: 1010400 meters

SUN POSITIONING:

... Azimuth: 64.8° (0-360°)

...Elevation: 90° (o-90°)

...Ambience: 0.7 (0-1)

by Mezemir Fikre-Mariam Wagaw, Feb. 2001

Institute of Geography

Faculty VII - Architecture Environment and Society

Technical University Berlin

3.1.2 The Quaternary Deposits and the Aden Volcanic Series

The pluvial sediments on the rift floor consist of lacustrine and fluviatile clays, gravels, diatomite, etc., while the interpluvial shows deposits of loess and aeolian sands. The main type of continental quaternary deposits found in the Ethiopian rift system is lacustrine. In the rift system, lava and pyroclastic deposits of the Aden volcanic series are frequently interbeded with these sediments.

For a better view of the rift valley and its escarpments, a rectangular north-south profile cutout with approximately 50 km width was taken from the USGS DEM, shown in [Figure 3] above. Using this, an image drape for 3D-model was done. The figures [Figure 4], [Figure 5] and [Figure 6] are DEM-models overlaid with the TM bands (4,3,2) in RGB, which includes the western escarpment, the rift valley and the eastern escarpment. The figures show the three main views on this profile from south, east and west side. It is important to note that the eastern lower escarpment and the western escarpment to a lesser extent, are covered - green biomass which shown as saturated red in the figures. The rift valley proper area is represented as a more bright and white, bluish to dark bluish color representing the uncovered reflective recent sedimentation and the morphological modulation respectively.

The Aden volcanic series is originated during Post-Miocene age. Most of these rocks are found in the rift system. The eruptions which gave rise to the Aden volcanic series, though numerous, are usually localized. It includes a great variety of lava types and many kinds of scoria and tuff. The dominant lava material is olivine basalt, which show a remarkable uniformity in its petrography even from flows of different ages or from different localities, [Pilger1976].

The pluvial lacustrine sediments of the rift system have tended to flatten out the rift floor to form extensive, monotonous flat plains as shown on a typical oblique picture in [Figure 7]. The volcanic cones are more clearly depicted in satellite imagery than on the geological map, as will be discussed in the next chapters. Numerous rhyolitic and trachytic domes are found in the rift valley especially in the central part. These volcanic domes and hills may act as groundwater barriers as they consist mainly of massive rocks with their often insulated intrusive structure.

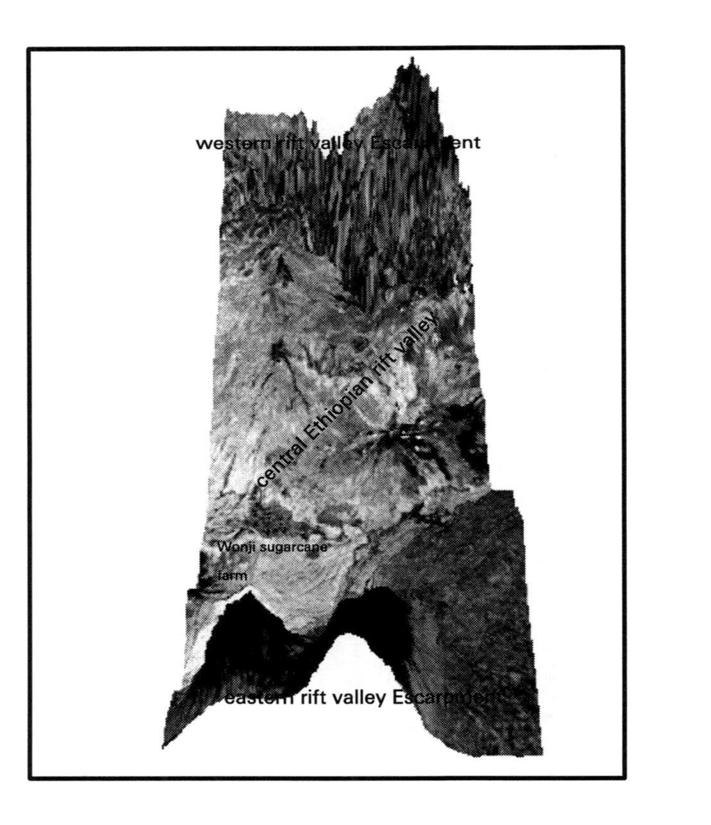

OBSERVATION POSITION:
...x = 545393 meters, y = 780254 meters
...AGL= 10000 meters, ASL = 10000 meters
..Direction:
...FOV: 50, Pitch = -48
...Azimuth: 0, Roll = 0.
SCENE PROPERTIES:
...DEM Exaggeration: 25
...Viewing Range: 325440 meters
SUN POSITIONING:
... Azimuth: 216° (0-360°)
...Elevation: 68.4° (o-90°)
...Ambience: 0.8 (0-1)

Figure 4. Observation from the south side on a north south profile cutout of the USGS DEM overlaid with a TM band combination 432 in RGB.

by Mezemir Fikre-Mariam Wagaw, Feb. 2001
Institute of Geography
Faculty VII - Architecture Environment and Society
Technical University Berlin

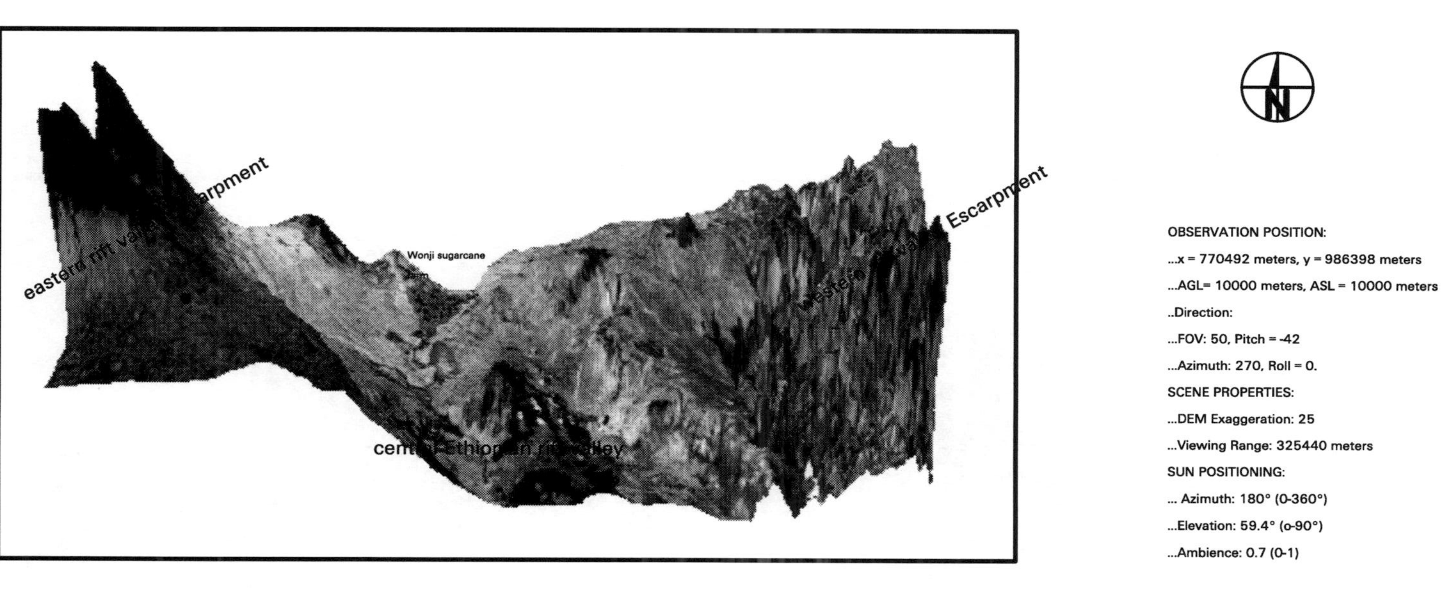

Figure 5. Observation from the east side on a north south profile cutout of the USGS DEM overlaid with a TM band combination 432 in RGB.

OBSERVATION POSITION:

...x = 770492 meters, y = 986398 meters

...AGL= 10000 meters, ASL = 10000 meters

..Direction:

...FOV: 50, Pitch = -42

...Azimuth: 270, Roll = 0.

SCENE PROPERTIES:

...DEM Exaggeration: 25

...Viewing Range: 325440 meters

SUN POSITIONING:

... Azimuth: 180° (0-360°)

...Elevation: 59.4° (o-90°)

...Ambience: 0.7 (0-1)

by Mezemir Fikre-Mariam Wagaw, Feb. 2001
Institute of Geography
Faculty VII - Architecture Environment and Society
Technical University Berlin

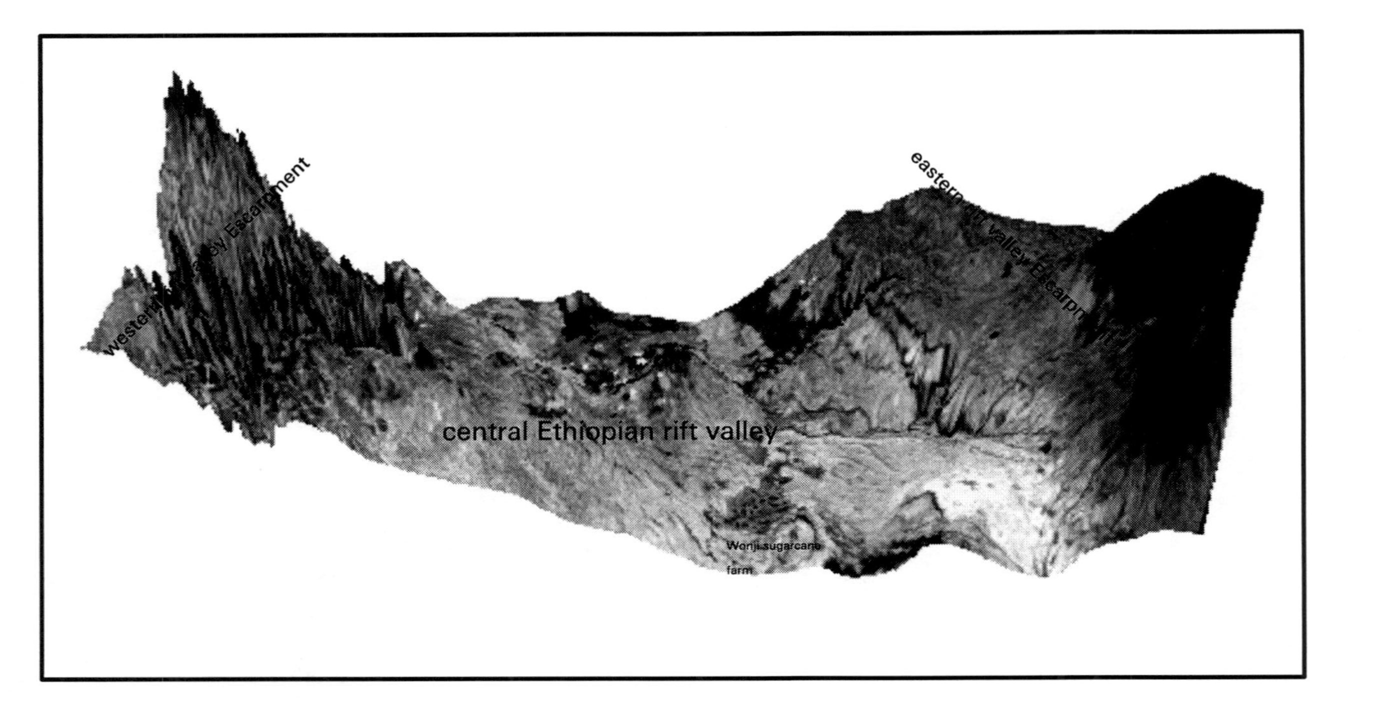

Figure 6. Observation from the westside on a north-south profile cutout of the USGS DEM overlaid with a TM band combination 432 in RGB.

OBSERVATION POSITION:

...x = 361097 meters, y = 925475 meters

...AGL= 10000 meters, ASL = 10000 meters

..Direction:

...FOV: 50, Pitch = -51

...Azimuth: 80, Roll = 0.

SCENE PROPERTIES:

...DEM Exaggeration: 25

...Viewing Range: 325440 meters

SUN POSITIONING:

... Azimuth: 180° (0-360°)

...Elevation: 59.4° (o-90°)

...Ambience: 0.6 (0-1)

by Mezemir Fikre-Mariam Wagaw, Feb. 2001

Institute of Geography

Faculty VII - Architecture Environment and Society

Technical University Berlin

Figure 7. Typical morphology of the rift valley area at three locations between the towns Wolenchiti and Metehara along the main road from Nazereth to Harar.

3.2 Climate of the Study Area

The climate in the study area is humid continental with hot humid summers and vigorous, warm sunny winters. The quantitative most beneficial rain, is essentially of orographic type, produced from condensation of vapors driven by winds against marginal escarpments on the plateau. In summer – mainly June to September - the rain is generally in the form of heavy downpour and slow shower, with sporadic thunderstorm. This rain season causes greater runoff. With less intensity, the same rain pattern occurs in late spring and early autumn. Much of the winter, end of October to February, is sunny, dry with a very little or no rainfall, see [Figure 8a].

Monthly mean average meteorological measurements for this study were available from the national meteorological service. This data was recorded at the Addis Ababa, Debre Zeit and Nazereth stations. It includes the monthly mean total rainfall and the maximum and minimum temperature data for the three locations. Average sunshine and pan evaporation were not available for Nazereth. Even though the available data is partially incomplete and the accuracy of measurements and measuring equipment is not commendable, it provides us with vital climatic information, which will be discussed below. In the discussion and analysis of the diagrams and tables in this sub-chapter, a word of caution that, the data is not always gap-free and therefore should be used with care.

3.2.1 The Addis Ababa Climate

As can be seen in [Table 2] below, the monthly mean annual average pan evaporation in Addis Ababa is 144 mm while the rainfall is just below 100 mm. This shows that even in the Addis Ababa area the pan evaporation is with 48% higher than the precipitation. The monthly mean yearly average maximum and minimum temperature of Addis Ababa are 22.6°C and 9.9°C, respectively. The annual average amount of sunshine is 6.7 hrs a day.

Table 2. Descriptive monthly mean yearly average statistical values for the Addis Ababa meteorological station.

statistical quantities	max. temp. (in °C)	min. temp. (in °C)	pan evapor. (in mm)	total rain (in mm)	sun shine (in hrs).
minimum	21.3	8.1	110.1	76.5	5.6
maximum	23.7	11.5	208.5	122.6	7.7
mean value	22.6	9.9	144.0	97.1	6.7
standard dev.	0.7	1.0	25.2	11.6	0.5
median	22.8	9.8	137.6	98.8	6.7
modus	23	10.4	110.1	76.9	6.6
span	2.4	3.4	98.4	46.1	2.1
skewness	-0.4	-0.1	1.1	-0.1	-0.2
kurtosis	-1.2	-1.1	0.9	-0.5	-0.1
number of samples	44.0	44.0	28.0	42.0	27.0

3.2.2 The Debre Zeit Climate

In the Debre Zeit area, the average annual mean maximum and minimum temperature is 26.4°C and 10.4°C respectively. The monthly mean yearly average pan evaporation and rainfall for the area is 145.8 mm and 66.4 mm respectively showing the rainfall to be only 45.5% of the pan evaporation. The average sunshine is 7.6 hours. The [Table 3] shows a summary information about Debre Zeit.

Table 3. Descriptive monthly mean yearly average statistical values for the Debre Zeit meteorological station.

statistical quantities	max. temp. (in °C)	min. temp. (in °C)	pan evapor. (in mm)	total rain (in mm)	sun shine (in hrs).
minimum	23.6	7.2	108.2	27.9	6.3
maximum	28.5	13.8	185.7	89.9	8.4
mean value	26.4	10.4	145.8	66.4	7.6
standard dev.	1.1	1.3	21.0	12.7	0.5
median	26.45	10.6	142.8	69.05	7.75
modus	26.5	11	108.2	75.9	7.9
span	4.9	6.6	77.5	62	2.1
skewness	-0.1	-0.1	0.2	-1.0	-1.0
kurtosis	1.8	1.5	-0.1	1.8	1.0
number of samples	24	28	22	30	16

3.2.3 The Nazereth Climate

As it is shown in the [Table 4], the average annual mean minimum and maximum temperature of Nazereth are 14°C and 27.9°C, respectively. With the monthly mean yearly average minimum value of 73.2 mm, the rainfall is slightly higher than the Debre Zeit area.

Table 4. Descriptive monthly mean yearly average statistical values for the Nazereth meteorological station.

statistical quantities	max. temp. (in °C)	min. temp. (in °C)	total rain (in mm)
minimum	25.3	12.4	36.6
maximum	30.0	15.4	108.6
mean value	27.9	14.0	73.2
standard deviation	1.2	0.8	18.7
median	27.95	14	70.7
modus	27.7	14	36.6
span	4.7	3	72
skewness	-0.3	-0.3	-0.1
kurtosis	-0.1	-0.6	-0.3
number of samples	28	31	33

3.3 Rainfall at the Three Stations

The main rainy season comes in all three locations simultaneously from June to September, with slightly varying peaks for Debre Zeit, at the end of June, and for Addis Ababa, in August. With 150 mm monthly mean rainfall for Addis Ababa from mid June to the end of September, and for Nazereth/Debre Zeit from the end of June to the beginning of September, the three stations get their maximum rain during these months. This main rainy season is crucial in replenishing fresh water reservoirs, spring reinforcements and groundwater regeneration.

From the end of February till the main rainy season, the monthly mean average precipitation is above 50 mm, [Figure 8]. This rain is tremendously important for the flora and fauna in the area as well as for providing some small-scale agricultural activities and grazing grass. In an average monthly comparison, Addis Ababa gets substantially more rain throughout the year, with its maxima in April (94.3 mm) also known as "Belg[i]" rain and August (280.4 mm) also known as "Meher[ii]" rain.

[i] Belg is an Amharic naming for the rain season in the months March, April and May.

[ii] Meher is an Amharic naming of the main rainy season in the months June to end of September.

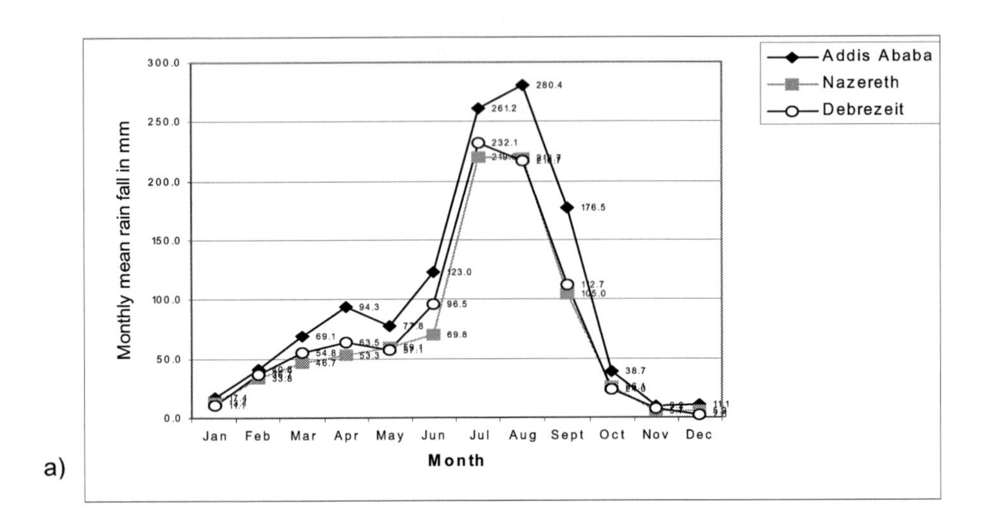

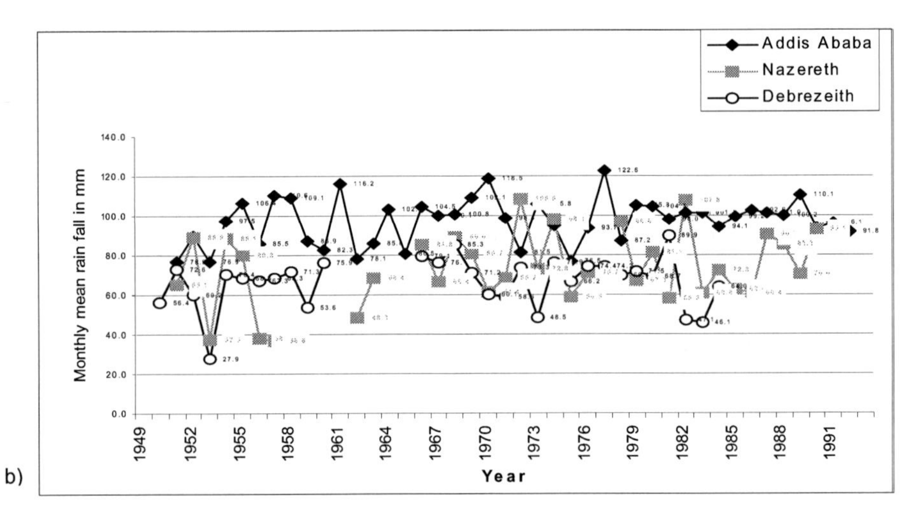

Figure 8. Monthly mean a) and monthly mean yearly average b) rainfall for Addis Ababa (1949-1993), Nazereth (1953-1993), and Debre Zeit (1958-1993).

In [Figure 8b] above, contrary to most expectations, there was a generally constant or slight increase of rain quantity in the course of the years from 1940 to 1993.

3.3.1 Mean Maximum and Minimum Temperature in the Three Stations

As shown in [Figure 9], the monthly mean maximum temperature of the three stations lies between 20°C and 30°C. It shows for Nazereth and Debre Zeit a high correlation. Between Addis Ababa and the other two locations, there is a constant mean maximum temperature difference of approximately 4°C. Throughout the measured years, the yearly average monthly mean maximum measurements tend to have a constant temperature value in Addis Ababa and this value of Addis Ababa shows far less fluctuation than that of Nazereth and Debre Zeit, [Figure 9b]. The maximum temperature for all the locations is in the month of May.

The average monthly mean minimum temperature for Debre Zeit asymptotes the Addis Ababa value in both monthly and yearly average values, [Figure 10].

a)

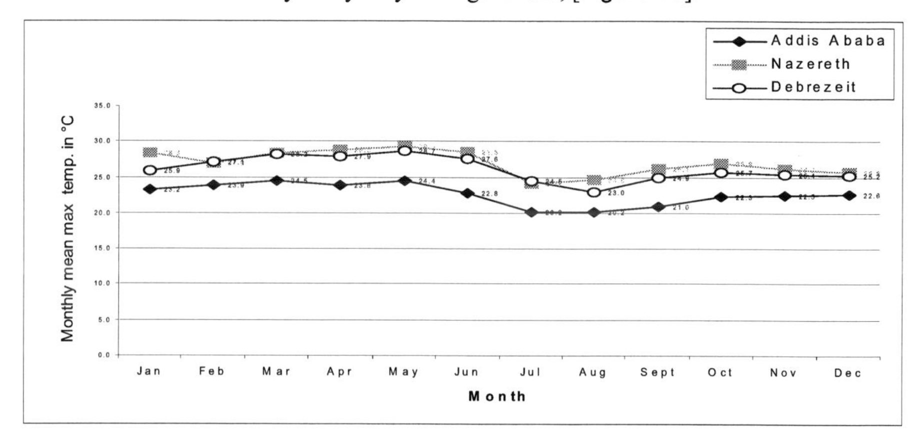

b)

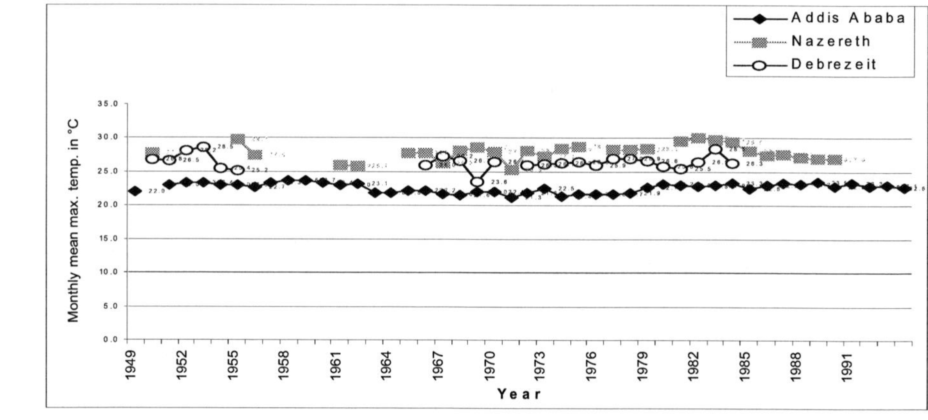

Figure 9. Monthly mean a) and monthly mean yearly average b) maximum temperature for Addis Ababa (1949-1993), Nazereth (1953-1993), and Debre Zeit (1958-1993).

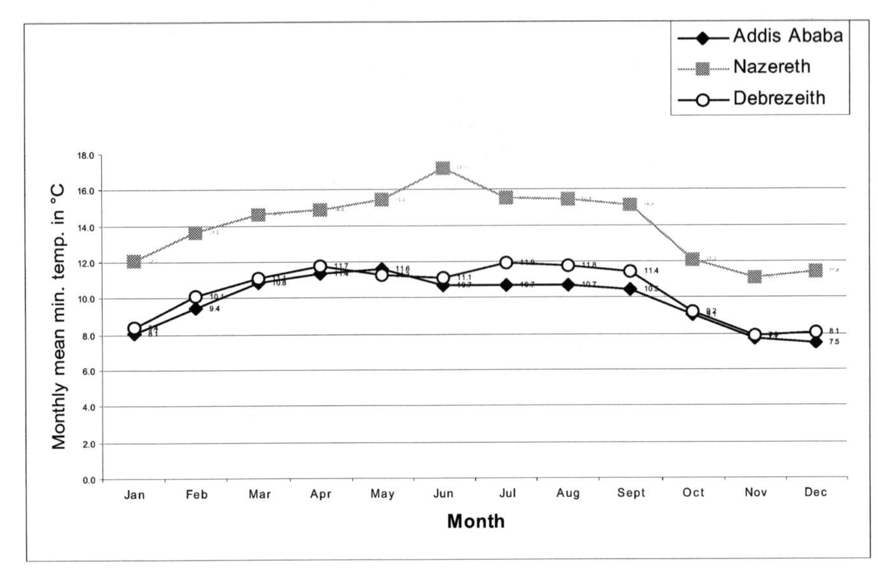

a)

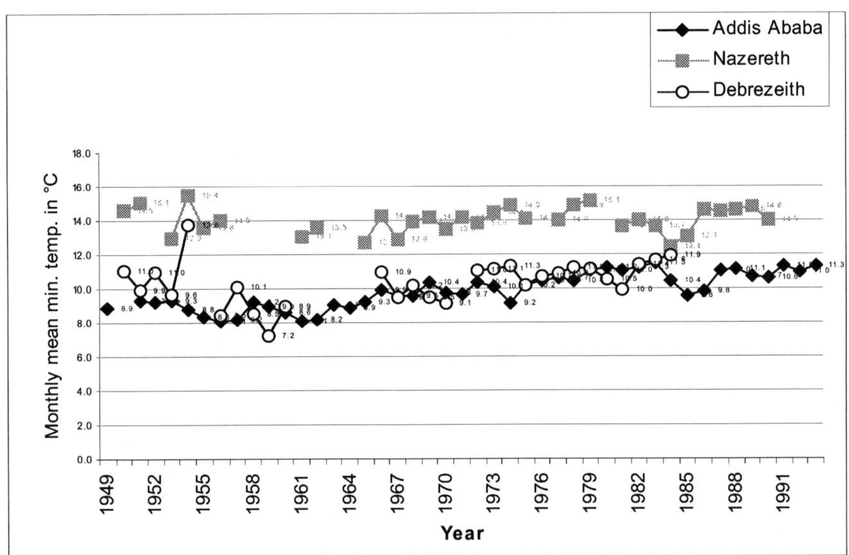

b)

Figure 10. Monthly mean a) and monthly mean yearly average b) minimum temperature for Addis Ababa (1949-1993), Nazereth (1953-1993), and Debre Zeit (1958-1993).

3.3.2 Sunshine in Addis Ababa and Debre Zeit

Except for the main rainy seasons of the year, June to end of September, the sunshine in Addis Ababa is over 6 hours per day. Even in the heavy rain seasons there is a minimum of 3 hours sunshine in a monthly average, see [Figure 11]. On the other hand, the yearly average monthly mean sunshine is 6.7 hours for Addis Ababa and 7.6 hours for Debre Zeit respectively. This figure depicts high availability of sunshine for a potential utilization of solar energy.

a)

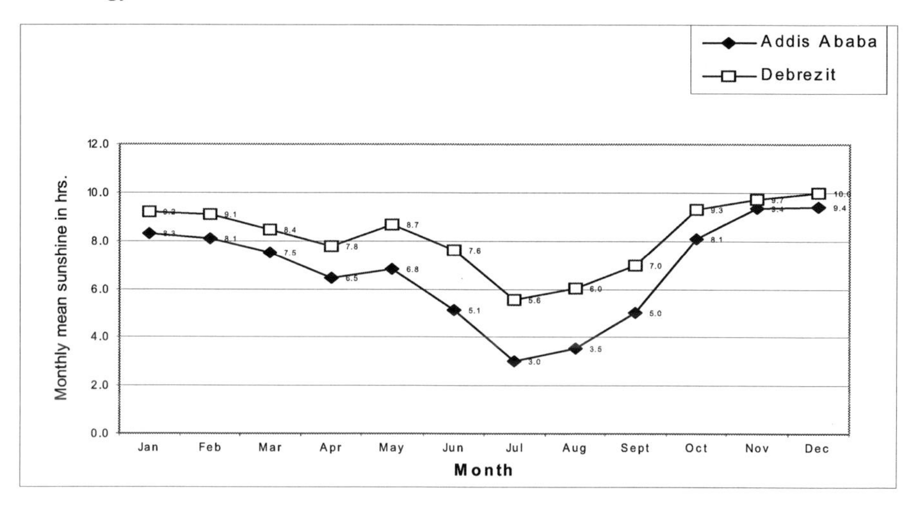

b)

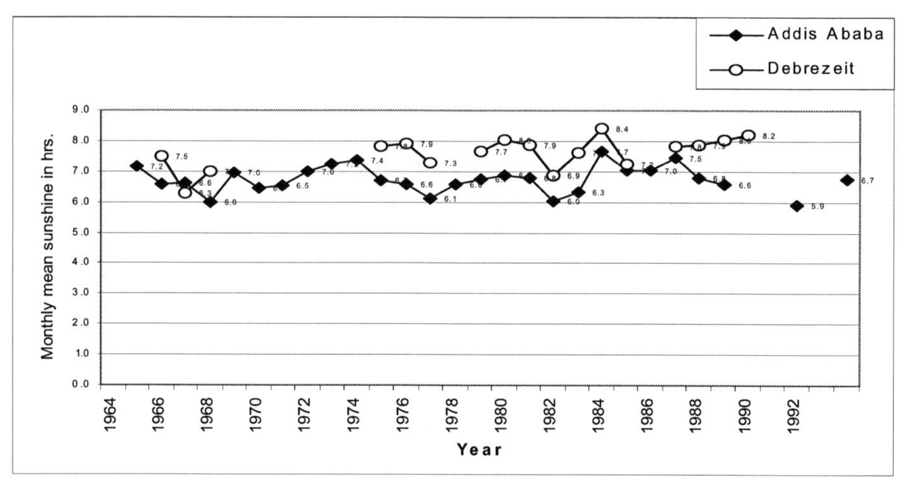

Figure 11. Monthly mean a) and monthly mean yearly average b) sunshine for Addis Ababa (1949-1993), and Debre Zeit (1958-1993).

3.3.3 Pan Evaporation in Addis Ababa and Debre Zeit

Even for the Addis Ababa area, where there is higher rainfall, the mean pan evaporation is higher in the months from October to May than the mean rainfall. Only for the four summer months June through September, the reverse is true, [Figure 12].

Throughout the year, Addis Ababa does show less mean pan evaporation than Debre Zeit except for the months of October through February, and the yearly average monthly mean pan evaporation in both locations shows a declining tendency over the measured years, [Figure 13].

a)

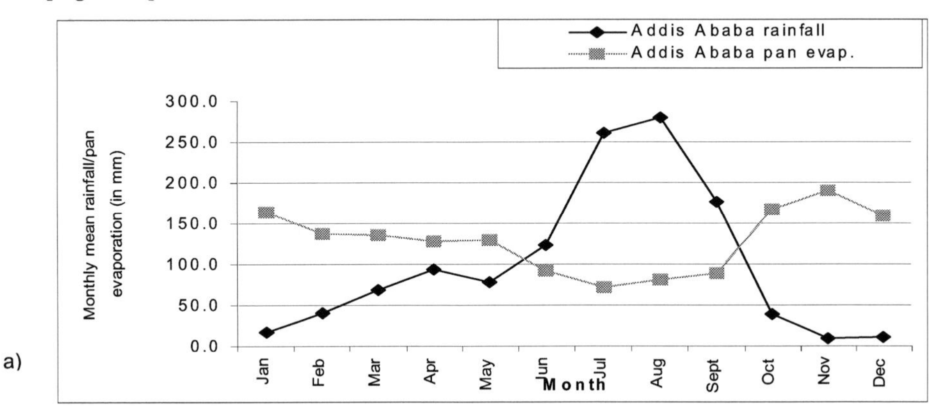

b)

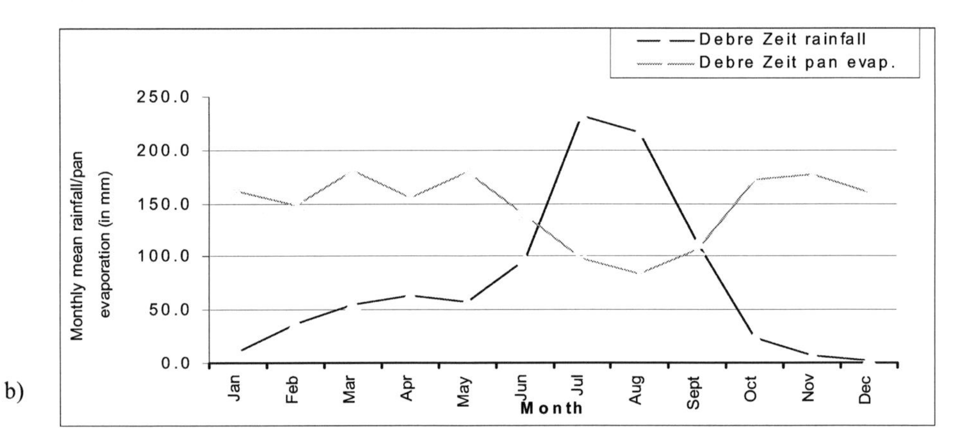

Figure 12. Monthly mean pan evaporation and rainfall in a) Addis Ababa for the years (1964-1993) and b) at Debre Zeit for the years (1966-1993).

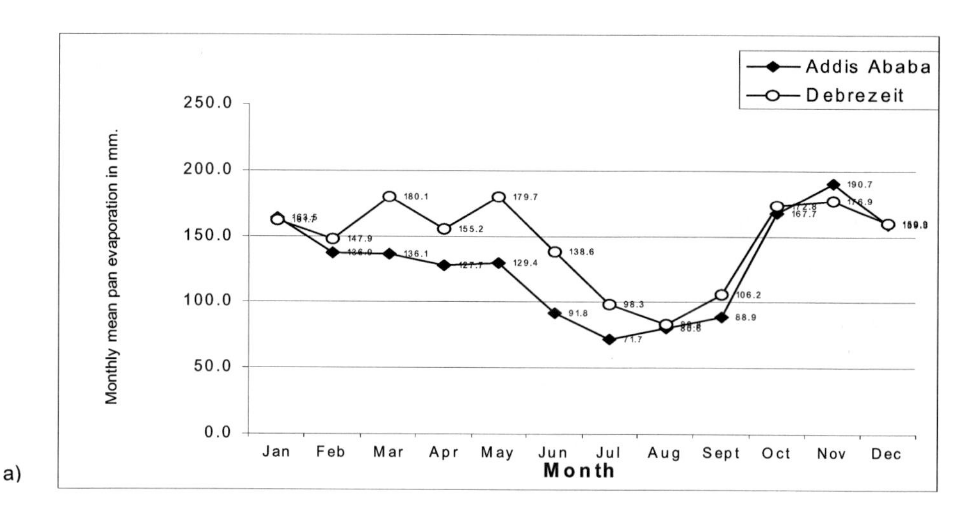

a)

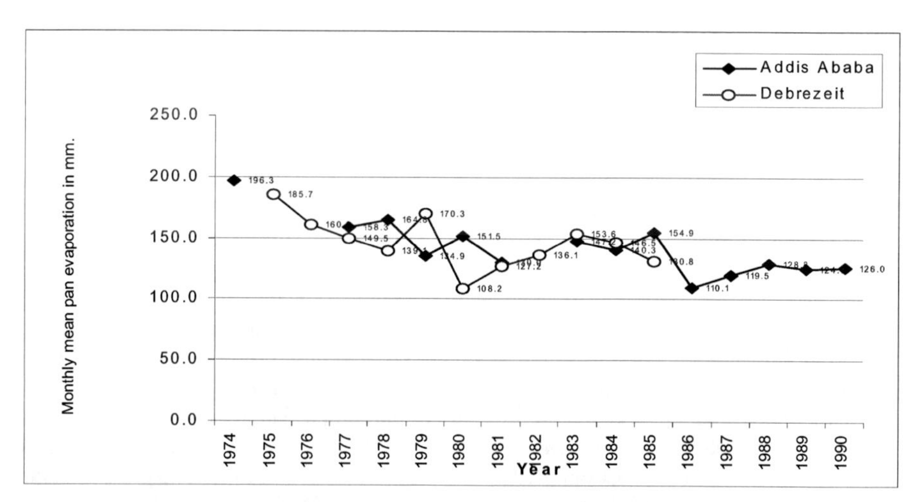

b)

Figure 13. Monthly mean a) and monthly mean yearly average b) pan evaporation for Addis Ababa (1949-1993) and Debre Zeit (1958-1993).

4 Remote Sensing Methodologies

4.1 Theoretical Review on Satellite Image and its Processing Methods

Remote sensing measurement and its subsequent utilization, generally require careful and knowledgeable interpretation and relies on a relatively large and robust data processing capability. The need for such capability results from the pressure, developed by the expanding information requirement associated with increased population, the availability of diverse data and a steady progress in enhancing the quality of life.

The qualitative and quantitative description and interpretation in a digital image processing is heavily dependent on the proper understanding of the spatial, spectral, radiometric and temporal properties of the sensed objects. Besides, knowledge of the signal theoretic models such as the signal transmission, filtering and enhancing methods are essential. In this chapter, a condensed literature study on the spectral characteristics of soil, vegetation and rock was given. The Landsat and SPOT imaging technologies and the different aspects of resolution and their modeling are presented. Different enhancement, restoration and error correction approaches are discussed. Vital image processing concepts which were relevant for this work, and their limitations are seen in detail. A flow diagram for the practical execution of the research was presented at the end.

4.2 The Landsat and SPOT Space born Sensors

The Landsat TM is composed of seven spectral bands, of 8-bit data with 30 meters of pixel size - with the exception of the band 6, a 120 meters coarse resolution - shows considerably higher spectral, spatial and radiometric resolution than the Landsat MSS with its 4 band of 6-bit data and approximately 79x56 meters pixel size, [Figure 14].

Three of the Landsat TM bands are in the spectral range of the Landsat MSS namely 0.52 to 0.60 μm (green), 0.63 to 0.69 μm (red) and 0.76 μm to 0.90 μm (reflective infrared). The green and red TM bands are narrower than the MSS which improved the sensitivity to spectral channels. The reflective infra red band is narrower than the combined bands of the MSS in this region, having its center in a region of maximum sensitivity to plant vigor. The blue (0.45 to 0.52 μm) band in the Landsat TM expands the use of satellite data in the field of bathymetry and stress evaluation for agricultural crops. The mid infrared bands (1.55 to 1.75 μm and 2.08 to 2.35 μm) will help in solving water-stress problems in crops as well as the geologists to better distinguish between rock classes. [Colwell1983] had summarized the principal application of Landsat TM as is shown in [Table 5].

*Table 5. Landsat Thematic Mapper functions and requirements after [*Colwell1983*].*

bands	spectral range (in µm)	radiometric resolution	principal applications
1	0.45-0.52	0.8% NEΔρ	coastal water mapping, soil/vegetation differentiation, deciduous/coniferous differentiation
2	0.52-0.6	0.8% NEΔρ	green reflectance by healthy vegetation
3	0.63-0.69	0.8% NEΔρ	chlorophyll absorption for plant species differentiation
4	0.76-0.9	0.8% NEΔρ	biomes surveys, water body delineation
5	1.55-1.75	0.8% NEΔρ	vegetation moisture measurement, snow/cloud differentiation
6	10.4-12.5	0.8% NEΔT	plant heat stress management, other thermal mapping
7	2.08-2.35	0.8% NEΔρ	hydrothermal mapping

With: NEΔρ representing radiometric sensitivity.

Using satellite technologies, the type of information acquired about the earth's surface are highly dependent on the spectral band of observation. Middle infrared sensors are sensitive to the surface thermal properties. Multispectral short and middle infrared sensors are sensitive to the chemical composition of the surface. All sensors are sensitive to the surface topography. No single sensor can provide a complete description of the surface properties. The main challenge in the interpretation task is to use the appropriate combination of sensors and/or spectral bands to acquire sufficient information using the physical, chemical and thermal characteristic properties of the object under investigation so that a reasonably good mapping for the requested environmental investigation is possible.

In the Landsat satellite system, sunlight reflected from the terrain is separated by a spectrometer into four or seven wavelength intervals, or spectral bands for the MSS and TM technologies, respectively. In the case of MSS, there are six detectors for each spectral band, thus for each sweep of the mirror, six scan lines are simultaneously generated for each of the four spectral bands. The detectors collect the energy and convert it into electrical signals for recording and transmission as image data.

In contrast to Landsat, the SPOT technology uses a line of sensors aligned perpendicular to the line of space craft motion. Its important characteristics are the fixed and equal detector-to-detector spacing, a long detector-integration time allowing very high signal/noise ratio, lightweight and its relatively small size, [Colwell1983_1].

For the SPOT, geometric processing is essentially the same as the other scanner technologies because earth rotation and panoramic projection effects will still occur. However, much of the along-line detector geometric rectification is not required.

4.3 Data Acquisition in Remote Sensing

The electromagnetic radiation, which comes from the sun to the earth and together with the partly emitted radiation back to the recording satellite, interacts with the atmosphere and any substance which it meets in the course of its propagation. Especially the high frequency part of the spectrum, and to some extent the whole electromagnetic radiation, is absorbed, reflected, refracted and diffracted by the atmosphere. For remote sensing purpose, only a limited atmospheric window is available which enables us to receive the - reflected and/or emitted - image signal, see [Figure 14].

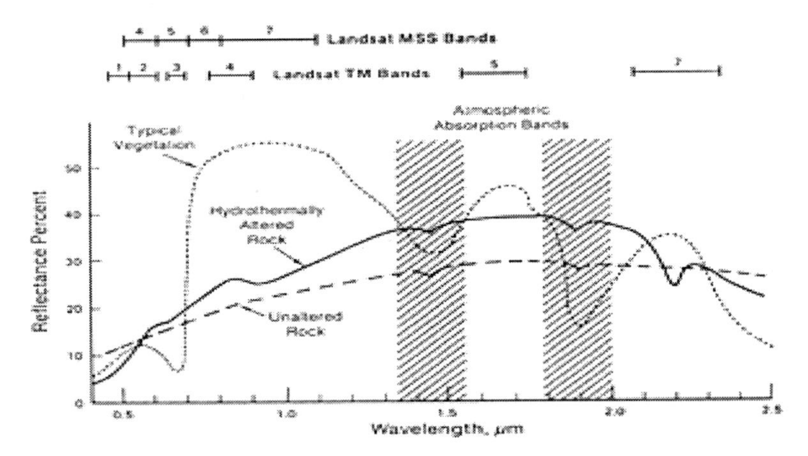

Figure 14. Generalized reflectance curve of green vegetation, superimposed on a diagram showing the spectral coverage of satellite sensing systems, reflectance curves for vegetation, unaltered rocks, and hydro thermally altered rocks (after [Sabins1983]).

Radiation emission may be a function of both view angle and wavelength. In addition, it is a function of the temperature of the emitting terrain element. This is usually taken to be the surface; however, it is also possible to have temperature gradients within the surface that could give rise to a volumetric emissivity. It is generally assumed that directional dependency of emissivity is minimum and that terrestrial emissivity is mainly caused by the heat energy i.e., the kinetic energy of random motion of the constituent particles of the earth's surface, [List1992_1].

On most thermal infra red images, the brightest tones represent the warmest radiant temperatures and the darkest tones represent the coolest temperatures. The thermal inertia of water is similar to that of soils and rocks, but on daytime, water bodies have a cooler surface temperature than soils and rocks. At night the relative surface temperatures are reversed so that water is warmer than soils and rocks, [Figure 15].

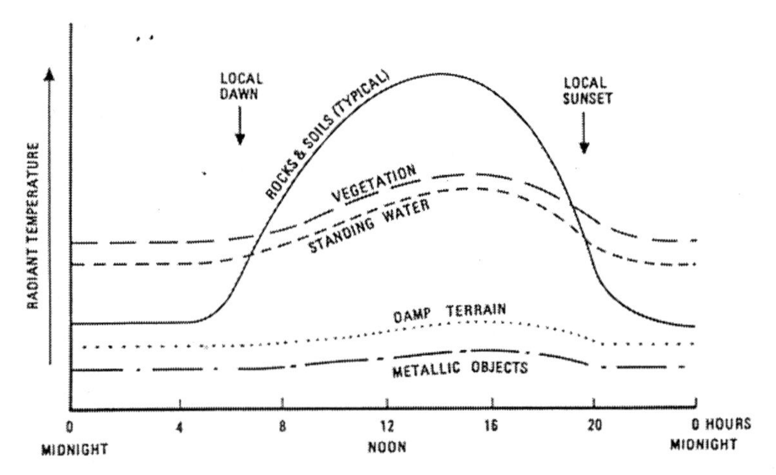

Figure 15. Diurnal radiant temperature curves (diagrammatic) for typical materials (after [Floyd1987]).

As shown in this diurnal temperature curve above, damp soil is cooler than dry soil, both at day and night. As absorbed water evaporates, it cools the soil. The effect of evaporative cooling dominates the radiant temperature signature of damp ground. Many geologic faults and fractures are recognizable in IR images because of evaporative cooling. Green deciduous vegetation have a cool signature on daytime images and warm signature on nighttime images. During the day, transpiration of water vapor lowers leaf temperature, causing vegetation to have a cool signature relative to the surrounding soil. Maximum reflectance from vegetation occurs in the photographic IR region, as shown by the bright tones in it, and in the reflected infrared region. These regions are mapped in the Landsat MSS band 6 and 7 as well as in the Landsat TM bands 4, 5 and 7. In addition, maximum spectral differences between vegetation types show up in the photographic IR region, that is advantageous for mapping plant communities. IR energy is almost totally absorbed by water, that causes water to have a dark tone on the IR images. The diurnal temperature range is a function of thermal inertia, which is directly related to density of the earth's surface materials and the surface moisture. Dense vegetation effectively mask other features on thermal IR images which lies between the vegetation. Moisture content is an important soil variable. Small changes in moisture content of the soil surface can produce significant changes in soil reflectance which is very important for this study.

4.3.1 Spectral Characteristics of Vegetation

In an infrared color combination, the red color of healthy vegetation is apparent, which is explained by the spectral reflectance curves of [Figure 14]. Different types of vegetation show a wider range of reflectance values in the photographic IR region than in the green wavelength region, which makes it easier to discriminate vegetation types on IR color than on normal color data.

The high infrared (IR) reflectance of leafs caused by the internal leaf tissue, or mesophyll, which consists of water-filled cells and numerous air spaces. The boundaries between these walls and air spaces strongly reflect IR energy. When the vegetation is stressed because of drought, diseases, insect infestation, or other factors that deprive the leafs water (moisture stress), the internal cell structure begins to collapse and the IR reflectance decreases. This decreased reflectance diminishes the red color on IR color photographs. The loss of IR reflectance is often called a previsual symptom of plant stress because it often occurs well ahead before the visible green color begins to change. Advanced plant stress produces a cyan color on a true color composite, [Kahle1981].

4.3.2 Reflectance and Emittance Spectra of Soil and Rock

Generally, soil reflectance curves exhibit a gentle increase with increasing wavelength. Soil reflectance depends upon the chemical and physical properties of the components, moisture content, iron oxide content, texture and surface roughness. Further, as organic matter content increases soil reflectance decreases, at least in the visible region. Generally soil reflectance increases for smaller particle size but this dependency is not always so straightforward in the field, where surface roughness and shadowing effects may predominate, [List1993_3].

Reflectance for rocks and minerals depends on both surface external effects, such as surface roughness and geometry, and internal causes related to their chemistry and internal geometry, giving rise to various molecular level processes. These processes include charge-transfer models, crystal field transitions, and transitions into the conduction band. Particularly with sedimentary rocks, spectral features may be associated with the constituents appearing as cements or impurities and in most cases reflectance is an indirect indication of rock composition. The short-wavelength infrared region 1 to 3 μm provides more diagnostic spectral information about the composition of minerals and rocks than the visible and near-infrared regions, [Goeth1981].

The mid-infrared region beyond 8 μm is especially important for geologic mapping because spectral emittance variations provide a basis for distinguishing between silicate and non-silicate rocks and for discriminating among silicate rocks. Some information concerning body properties as opposed to surface properties was obtained by analyzing the changes in surface temperature that are included by diurnal solar heating. This method allows measurement to a depth of about 10 cm or less, [Gillespie1984].

Thermal infra red images records the radiant temperature of materials. Radiant temperature is determined by a material's kinetic temperature and by its emissivity, which is a measure of its ability to radiate and to absorb thermal energy. The diurnal temperature change is a function of thermal inertia, which is directly related to density of materials. Thermal IR images are useful for many applications, including differentiation of rock types. Denser rocks, such as basalt and sandstone, have higher thermal inertia than less dense rocks, such as cinders and pumice and silt stone, [Kahle1984].

4.3.3 Spatial Resolution and the Instantaneous Field of View

As sensing aperture sweeps across the scene, it will filter the scene spatial frequencies. The spatial resolution, in terms of the geometric properties of the imaging system, is usually described as the instantaneous field of view (IFOV). IFOV is a function of satellite orbital altitude, detector size, and the focal length of the optical system.

An IFOV value is not in all cases a true indication of the size of the smallest object that can be detected. An object of sufficient contrast with respect to its background, can change the overall radiance of a given pixel so that the object becomes detectable.

The [Table 6] shows that very different estimates of resolution can be obtained for the same sensor. The choice of the IFOV depends mainly on which image properties the user is interested in as determined not only by the application but also by the method of analysis.

The spatial resolution of a system must be appropriate if one is to discern and analyze the phenomena of interest. The phenomena of interest may be natural or cultural features, and these may coexist within the environment at macro-, meso- or micro-scales. Each increase in scale requires progressively finer resolution of data in order identification and analysis to be successfully accomplished. Since image information content is resolution-dependent, a trade-off exists between the levels of resolution for most remote sensing applications. High spatial resolutions provide small area observation, in this case regional patterns may be difficult to characterize. Low or coarse spatial resolution allows regional patterns to be readily observed, interpreted and analyzed. However, detail may be averaged within a pixel, implying a loss of information. Every measuring system is sub-optimal only for certain objects in a scene.

A resolution employing spectral properties of the target is the Effective Resolution Element (ERE). The ERE is that part of a scene, which is contributing to a single digital measurement, widely known as Digital Number, DN. It is primarily defined by IFOV, but is also affected by the optics, electronics and sample spacing in the instrument. It may be further affected by image-motion smear and the variable effect of the atmosphere. [Colvocoresses1975] defines an ERE as the size of an area for which a single radiance value can be assigned with a reasonable assurance that the response is within 5% of the value representing the actual relative radiance.

From the preceding discussion of technically derived measures of spatial resolution, it is clear that no single definition is satisfactory. For gaining maximal information about a given geographical area, the analysis of images with different geometrical resolving power could give a vital clue of information. Basically, it is possible to produce from a given single image different scale maps. The production of the different scale maps is helpful in the interpretation and analysis stage, due to the scale of information inherited in them.

In a typical human interpretation, to move from detection to identification, the spatial resolution must be improved by about 3 to 10 times. About ten to one hundred times resolution-increase is necessary to pass from identification to analysis, [Simonett1976].

Table 6. Summary of estimates of the resolving power that have been calculated for Landsat MSS (after [Townshend1980]).

resolution measure	source	resolution (in meters)
1. IFOV-geometric	NASA (1972)	79
2. IFOV-geometric	Slater (1979)	76.2
3. IFOV-geometric	Colvericoresses (1979)	73.4
4. IFOV-geometric (min. altitude)	Gordon (1980)	76
5. IFOV-geometric (max. altitude.)	Gordon (1980)	81
6. pixel size	General electric (undated)	79*56
7. pixel size-resampled (Landsat 3 CCT's)	Holkenbrink (1978)	57*57
8. IFOV-point spread	Landgrebe, et. al. (1977)	90
9. EIFOV-half cycle	Welch (1977)	66
10. EIFOV-full cycle	Welch (1977)	135
11. ERE	Colvocoresses (1979b)	87
12. modified ERE-estimate for channel 4	Norwood (1974)	125
13. minimum classifiable area	Shay et al. (1975) General Electric (1975)	500*350 320*220

4.3.4 Spectral and Radiometric Resolution

Spectral resolution pertains to the width of the regions in the electromagnetic spectrum that are sensed and the number of channels. It entails sampling the spatially segmented image in different spectral intervals. The spatial resolution can modify the spectral signature, which implies that the spatial resolution provides a form of spectral filtering [Slater1983].

The spectral resolution of a remote sensing instrument is determined by the bandwidths of the channels used. High spectral resolution, thus, is achieved by narrow bandwidths which, collectively, are likely to provide a more accurate spectral signature for discrete objects than is the case for broad bandwidths. In contrast, broadband sensors usually have good spatial and radiometric resolution. This may be alleviated if relatively long look (or dwell) times are used during imaging [Lintz1976].

A radiometric resolution is the sensitivity to differences in signal strength. It is strongly related to the identification of scene objects, the finer the resolution the greater the opportunity to discriminate between objects commensurate with a given spatial resolution. It can also be understood as the resolving power of an instrument needed for dividing the total range (from black to white) of the signal output. The greater the number of discriminable grey steps - or resolved signal levels - the more radiometrically resolved are the collected data. Radiometric resolution is determined by the number of discrete levels into which a signal may be divided. The incrementing of the output leads to a reasonable number of just-discriminable levels. The increment is thus a function of the uncertainties introduced by the noises of the environment, the sensor, and the signal encoding method, [Slater1983].

Within a given spatial resolution, increasing the number of the discretion levels or improving the radiometric resolution will improve discrimination between scene objects. Interdependencies between spatial, spectral, and radiometric resolutions for each remote sensing application affects the various trade-offs. These trade-offs will be governed by the particular application or group of applications envisaged to solve using these data and the contrast among the sensed objects. There is a parallel in all the above discussed resolution types. In each we are concerned about the variability of the environment within the sample window. Additionally the appropriate dimensional spacing of samples in the direction of an increase or decrease of the dimensional unit such that significant environmental changes are retained. In each case, we find the Nyquist criteria setting the limit. The Nyquist interval-limit requires a minimum of two samples per repetitive cycle as the sensitivity to differences in signal strength, [Lintz1976].

4.3.5 Geometric Distortion and their Modeling

Geometric distortions can be categorized into sensor-related distortions such as aberrations in the optical system, non linearity and noise in the scan deflection system, sensor platform-related distortions caused by changes in attitude and altitude of the sensor, and object-related distortions caused by earth rotation, curvature and the terrain relief. Perspective distortions depend on the type of sensor used [Moik1980].

The basic requirement for geometric correction processing is the acquisition of control points in the image to be corrected. The coordinates of points in the image and corresponding points in a reference are inputs to the distortion modeling process by which an image is ultimately corrected. Control point location accuracy measures the geometric fidelity of the map i.e., how accurately the position of a known point on the ground is represented on the map. These properties are specified by the scale of the map and the projection to which it conforms. With this information, the geographic coordinates of a point on the earth's spheroid surface will be mathematically transformed to the coordinates on the map.

As one task part of a digital image processing, the elements of a digital image have to be transformed into a universal geographic coordinate system. After such transformation, it will be possible to compare multi-sensor data and the transformation of information from topographic and other maps. The purpose of geometric correction is to locate image samples such that the locations are known with respect to a map-grid reference system (geodetic rectification) and each satellite-pass over an area have to be digitally registered. Geometric transformations are used to correct the geometric distortions, make an overlay of images on other images (image registration), correct the aspect ratio, rotate the skew and flip images [Moik1980].

4.3.6 Radiometric Degradation and their Restoration

A pixel is a result of superimposition of different reflecting/emitting objects of a given square area on the surface of the earth that will be agglomerated with the sensor's Instantaneous Field of View (IFV). The grey level of each of the bands in the satellite pictures is the result of this process. In a detailed account, the spectral response patterns from the surface categories as well as others vary due to the natural random variations, systematic seasonal causes, atmospheric haze, etc. There is no unique measurement pattern associated with each category. Rather, associated with each category is a probability distribution indicating, for any measurement pattern, the relative frequency of occurrences that may arise from a ground area of the given category.

Random-noise image degradations are assumed to be homogeneous, random processes uncorrected with the signal, and are not amenable to restoration in this investigation. The main types of coherent noise appearing in images are periodic, striping, and spike noise [Moik1980]. Only generic forms of image enhancement - i.e. forms that use local pixel-neighborhood averaging to reduce the noise with some sacrifice of image edge details – can be readily applied. The radiometry or intensity of the sensor output may be changed to correct the data, to match its output with the display or recording devices, or merge diverse data sets.

Absorption and scattering by the atmosphere between a scene or target and a remote sensor results in a contrast reduction of all the elements in the scene. In most cases, the effect on system performance is adequately represented by assuming a constant contrast reduction factor, which is independent of spatial frequency. Motions of the image during the exposure time result in a smear, which degrades the image. These motions may be linear, such as result from uncompensated forward motion of the vehicle, sinusoidal, such as result from vibrations or may take other forms.

Radiometric degradation arise from blurring effects of the imaging system, nonlinear amplitude response, vignette and shading, transmission noise, atmospheric interference (e.g., scattering, attenuation, and haze), variable surface illumination (e.g., differences in terrain slop and aspect), and changes of terrain radiance with viewing angle. It causes point (pixel) and spatial image degradations. Pixel degradations occur in some imagery systems when object brightness is not uniformly mapped onto the image plane for example due to vignette. The response of the system to a point light-source is described by the point spread-function. Radiometric spatial degradations are caused by defocusing, diffraction effects, and by atmospheric turbulence and relative motion between the imaging system and the object. Radiometric degradations caused by coherent noise are amenable to restoration. The essence of coherent noise removal is to isolate and remove the identifiable and characterizable noise components in a manner that it does a minimum of damage to the actual image data. Different techniques may be applied in order to minimize these degradations. For the atmospheric correction [Orthaber1999] have developed a method using at least one visible short wave (blue) and near infrared spectral band (e.g. of Landsat TM) for regions like the Alpine which are covered with relatively dense vegetation.

4.4 Applied Digital data Enhancement and Noise Filtering Methods

The presence of noise in the measurement process leads to a measured image that differs from the true scene by the error introduced by the noise. For purpose of simplification noise can be divided into two broad categories, namely the signal-independent noise and signal-dependent noise. For signal independent noise there is a component added to the image that comes from the sensor, the transmission path or some other source. The complete characterization of a noise process requires a statistical description that provides the probability density-functions of single and multiple samples taken from the data collection process. Generally such detailed information is not available and some assumptions must be made that permit the processing to be carried out without danger of serious degradation of the image. Often in a sampled and quantized data, the probability density function of the quantization noise is uniformly distributed over the quantizing increment and the frequency spectrum is constant to those frequencies with values many times that of the sampling frequency. Most image processing technologies are designed on the assumption that the noise is uncorrected from pixel to pixel and has a uniform frequency spectrum. Noise can be characterized by its effect on the probability that a given digital sample will be assigned the precise value corresponding to the true scene radiance at that point [Billingsley1975].

In a theoretical noiseless system there is no ambiguity in the designation of a particular signal level as a certain digital number (DN). However, in the presence of noise (assumed to be random), it is the signal and the noise which are quantized, and the level of the signal alone is somewhat uncertain from inspection of the digital number. Specifically, there is a finite probability measured by the relative area of the probability distribution bounded by the quantization step boundaries.

There are different digital filtering methods and tools for removing noises from the image which are generated at different stages of data collection and transmission process. These tools are equally relevant for splitting and enhancing the targeted information form the original image that inherently contains much more information content than it is generally needed by a specific application and analysis request.

4.4.1 Filtering Using the Moving Average - the Convolution Matrix

The spatial information in an image can be considered as being composed of low and high frequencies. The low frequency component is usually represented by large areas with constant brightness, which in the case of satellite image data, usually represents the albedo color information. High frequency information consists brightness changes over a short spatial dimension that occur because of contrast in slope attitude or topographic features or contrast in brightness at boundaries between different units of the geology.

The digital spatial filtering technique also known as convolution filter works in image space and gives equal weight to every picture element (pixel) within the filter window or kernel which is used. The high frequencies or details that are enhanced can vary from very fine information that is close to the resolution limit of the mapping system to the structural information hundreds of pixels long. The size of the kernel used is determined by the size of the linear features we are interested in mapping. A spatial filter generally enhances features that are half the size of the kernel being used and suppress features that are more than half the kernel size. Kernel sizes will be kept equi-dimensional to ensure that all directions would be weighted equally. The kernel size that was used for edge enhancement and lineament detection may be selected based on the number of edges or how "busy" the satellite image is. This type of kernel also helps to reduce linear artifacts that can be seen in the directional directional-filtered images as described by [Gillespie1976].

A wide variety of approach as are possible to achieve edge enhancement. The filters used in these approaches are either developed from models of image-degradation effects or chosen on a heuristic basis because they improve the visual presentation of the data. The maps generated by the horizontal first difference technique 31x31, 51x51 and 101x101 convolution matrix, as will be seen in the next chapter, provides to be an excellent data source for obtaining the study area's structural fabric.

4.4.2 Multispectral Image Transforms- the Principal Component Analysis

Image transforms results in a reduction of data-set size while retaining the essence of the extractable information. The Principal Component Analysis, PC, also known as eigenvector transform, is a standard statistical technique for selecting that subspace of given dimension in which the most data variance lies.

A variety of linear transformations are available to implement transform coding on satellite data, but the Principal Component-approach is the optimum linear transformation due to the fact that the mean square error between the original and reconstructed data is less than for any other linear transformation, and it eliminates or suppresses all correlation in the data. Thus the principal component transformation is useful for better aggregation of the information.

The stretched inverse-PC was applied for generating best suited visual interpretation material, because it gives back the detail of each grey level especially in the extreme white and extreme black parts of each band as will be seen later on in the next chapter.

4.4.3 Ratioing and Intensity Hue Saturation (IHS) Transformation

Ratioed images mainly suppress brightness variations due to topographic relief and enhance subtle spectral variations. Ratioing may concurrently increase random or coherent noise or atmospheric effects. Therefore, coherent noise (e.g., striping) shall be removed before ratioing and differencing. In this processing two or more images could be subtracted to emphasize spectral changes. Subsequent rescaling (contrast stretch) is performed to make the range of the difference or ratio images compatible with the dynamic range of the display device [Moik1980].

The IHS-system is a polar cylinder coordinate system and is another way of describing the color combination of a given image. The saturation and the intensity represents the x- and z-axis respectively. The hue is represented as the angular axis with respect to the saturation axis. The owned colors in the IHS domain do not directly represent the natural or near infrared color. But pictures with such a representation are very relevant for the visual interpretation and classification.

Using this, it is possible to generate and manipulate different data in each channels of the IHS domain, often the hue is substituted with geological maps, digital elevations, etc. The IHS transformation is also useful for combining SPOT and TM images in order to become a high geometrical and spectral resolution output. With the help of the geometrical rectification the two different resolution image data can be transformed on each other. The I-channel in the IHS domain shows the intensity distribution within the color image and it is compatible with the panchromatic SPOT channel. By substituting the I-channel with the SPOT image and back transformation to the RGB domain, one becomes the high-resolution part of the SPOT-image. In this process, the grey values of the SPOT image will manifest themselves as a varying brightness also within the TM-pixels. As a result high geometrical resolution from the SPOT image will be gained without much loss in the spectral aspect. This substitution of the intensity is inherently justified, under the condition that the intensity-channel from the TM represents a panchromatic characteristics, i.e., when this intensity is from the spectral bands which lies approximately to that of the spot panchromatic spectral characteristics. This is mainly the case when the TM (1,2,3) bands are displayed in RGB respectively.

4.5 Workflow for the Practical Image Interpretation

In the above discussions we have seen remote sensing data collection concepts, their sampling, error sources and different processing methods. For this particular study, the image processing methods, which were discussed up to now, are compiled as a flow diagram, [Figure 16]. The results in the next chapter were achieved using the Erdas Imagine[iii] software version 8.3.1 and the flow diagram below.

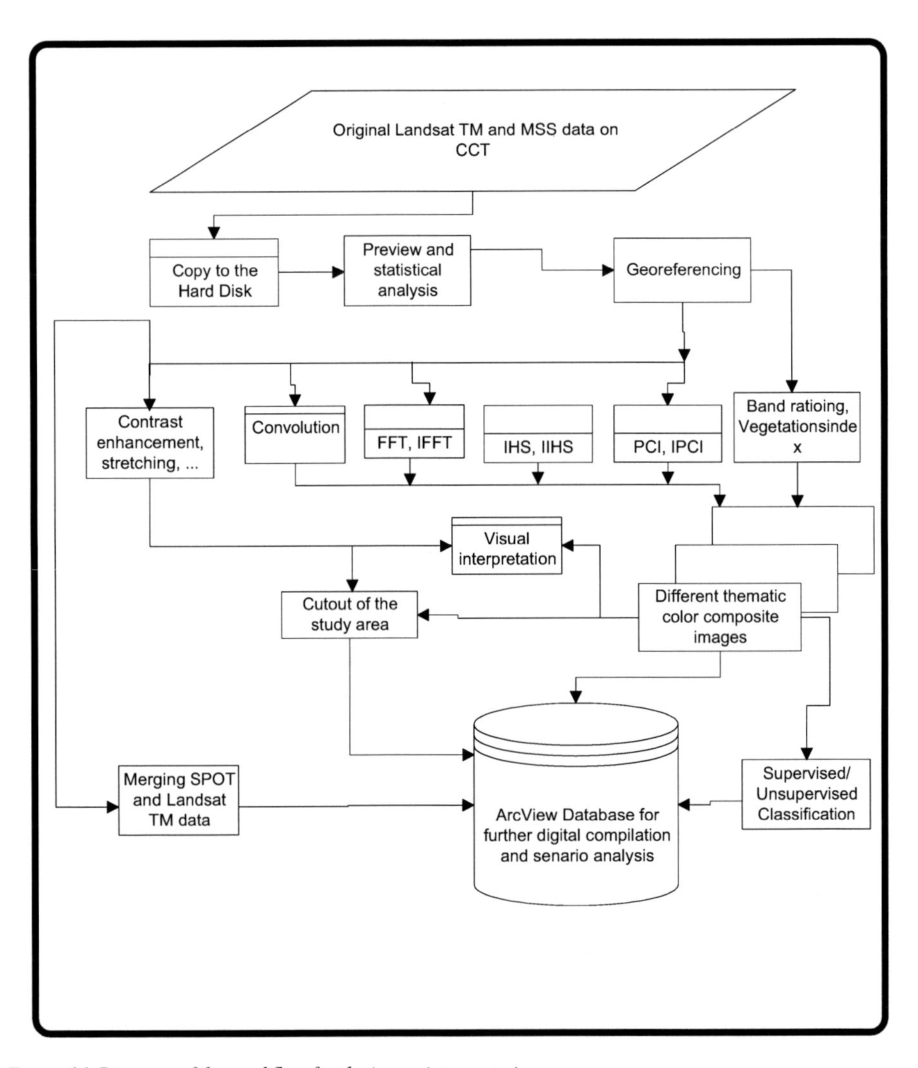

Figure 16. Diagram of the workflow for the image interpretation.

[iii] Erdas Imagine is the product of ERDAS Inc.

5 Results of the Image Processing and Discussion

In the past chapters, the geology and geomorphology of the study area was introduced, the theoretical background of signal theory and remote sensing discussed, and flow diagram for its practical implementation prepared. In this chapter the available data will be processed, interpreted and a discussion about the results will follow. In the first part of this chapter, the statistical quantities from the whole scene will be described. Then after cutting out the study area from the whole scene, different operations such as PC analysis, IPC, contrast stretching, IHS, inverted IHS analysis, convolution, FFT and IFFT operations will be performed. Results will be discussed. In order to extract the regional structure of the area, best-fitting convolution matrix is selected and applied. Different band rationing, differencing and 100% feedback operations are done. A rose diagram for the lineaments in the area of consideration is drawn. Finally a comparative discussion between the available hydro geological/geological map cutouts and the acquired result was done.

5.1 Grey Level Digital Number-Value Distribution of the Data

The Digital Number, DN, value distribution of each imagery had given a vital clue about the information content of each of the spectral bands. The form of each histogram shows the extent of each scene and allows to draw information about the homogeneity and normal distribution of the data. Additionally it gives information about the properties of the scene. A scene with a homogenous surface and less contrast gives a histogram with a single and distinct mode and those with less contrast resolution or amount shows a skewed and multi-mode diagram. Generally, if the scene gives a single and wide mode, then the histogram represents a homogenous image with a higher contrast. As shown in the [Figure 17a] of the Landsat MSS histogram below, the spectral bands 5, 6 and 7 are highly correlated. In [Figure 17b] of the Landsat TM, the seven bands show a correlation, even though it is in a lesser extent, for some of them.

Basically, the design of the imaging sensor is in such a way that it can differentiate and capture a very wide range of signals from the target in the given respective spectral bands. Due to this, within a satellite scene, the relevant signals are confined in a relatively small dynamic range. If the grey levels are all clustered together, the image will have very low contrast. This low contrast will be, at least partly, restored by using an appropriate software handling on the input data. This handling includes such operations as histogram linearization and stretching. They allow the use of the whole dynamic range efficiently.

The form of the grey level distribution against the number of grey level values shows the resolution contrast of each band. The correlation analysis of the histogram gives us an indication of the information content and information redundancy across the whole bands. Especially the neighboring bands of a multispectral image are generally highly correlated. The tables [Table 7] and [Table 8] summarize the descriptive statistical information, variance-covariance, and correlation matrix of the Landsat TM and MSS data.

In the scatterogram of the original Landsat MSS and TM image, see [appendix 1] and [appendix 2] respectively, the presence of most of the data along the diagonal indicates a high degree of correlation between the luminescence and emittance of points separated by the sampling distance used. As the separation of the points in the scene increases, the joint probability density-function spreads out until there is a noticeable lack of data along the diagonal, signifying that points separated by large distances are most likely not equal. For the MSS these diagrams show more clearly the linear dependency of band 6 with 7, and also in a lesser degree bands 5 with 6 and 5 with 7. The combination of the band 4 with all other results to convey maximum information. The scatterogram for TM shows band 3 with band 7 and band 4 with band 7 to be maximally correlated. The most uncorrelated band is band 6.

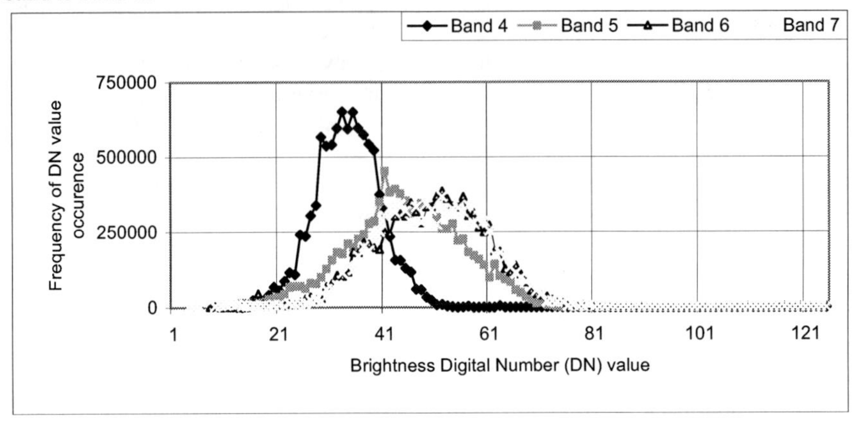

a)

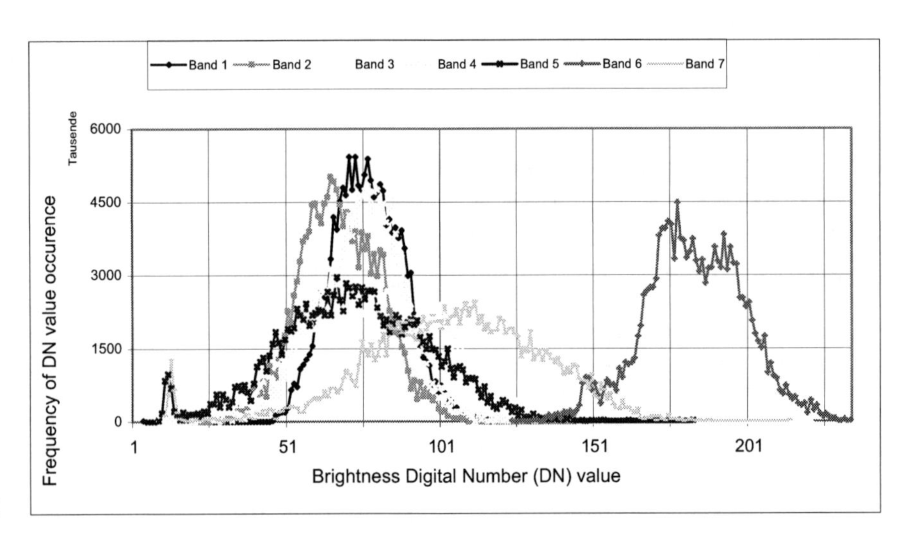

b)

Figure 17. Grey-level value distribution of the unprocessed Landsat a) MSS image with 3168 rows x 3161 columns x 4 bands and b) TM image with 5373 rows x 5066 columns x 7 bands. The form of the grey value distribution shows the resolution contrast of each band and the intensity of correlation with each other.

Table 7. *Statistical values of the Landsat MSS scene composed of four bands each with 3168 rows x 3146 columns of pixels.*

variance and covariance of the MSS				
	band 4	band 5	band 6	band 7
band 4	2.92E+10	9.93E+09	4.77E+09	4.71E+09
band 5	9.93E+09	1.40E+10	1.26E+10	1.28E+10
band 6	4.77E+09	1.26E+10	1.41E+10	1.39E+10
band 7	4.71E+09	1.28E+10	1.39E+10	1.40E+10
correlation coefficient of the MSS				
band 4	1			
band 5	0.49	1		
band 6	0.23	0.91	1	
band 7	0.23	0.91	0.98	1
descriptive statistical information of the MSS				
art. mean	7.68E+09	7.68E+04	7.68E+09	7.68E+09
std. dev.	1.72E+09	1.19E+09	1.19E+09	1.19E+08
minimum	0	0	0	0
maximum	6.50E+05	4.54E+05	3.87E+05	3.69E+05
median	0	2048	4096	7168
mode	0	0	0	0
samples	127	127	127	127

Table 8. *Statistical values of the Landsat TM scene composed of seven bands each with 6525 rows x 7160 columns of pixels.*

variance and covariance of the TM							
	band 1	band 2	band 3	band 4	band 5	band 6	band 7
band 1	9.83E+10	-1.76E+10	1.58E+10	2.54E+10	2.05E+10	-2.87E+10	-8.09E+07
band 2	-1.76E+10	1.46E+11	2.66E+10	2.03E+09	-1.41E+10	-2.94E+10	4.70E+10
band 3	1.58E+10	2.66E+10	6.63E+10	6.45E+10	-3.52E+09	-2.92E+10	5.12E+10
band 4	2.54E+10	2.03E+09	-3.52E+09	7.83E+10	-9.22E+08	-2.91E+10	4.74E+10
band 5	2.05E+10	-2.94E+10	-2.92E+10	-9.22E+08	3.04E+10	-6.54E+09	-7.90E+09
band 6	-2.87E+10	-2.94E+10	-2.92E+10	-2.91E+10	-6.54E+09	1.53E+11	-2.90E+10
band 7	-8.09E+07	4.70E+10	5.12E+10	4.74E+10	-7.90E+09	-2.90E+10	6.01E+10
correlation coefficient of the TM							
	band 1	band 2	band 3	band 4	band 5	band 6	band 7
band 1	1.00E+00						
band 2	-1.47E-01	1.00E+00					
band 3	1.95E-01	2.71E-01	1.00E+00				
band 4	2.90E-01	1.90E-02	8.95E-01	1.00E+00			
band 5	3.74E-01	-2.11E-01	-7.83E-02	-1.89E-02	1.00E+00		
band 6	-2.12E-01	-1.77E-01	-2.66E-01	-2.43E-01	-9.27E-02	1.00E+00	
band 7	-1.05E-03	5.02E-01	8.11E-01	6.91E-01	-1.85E-01	-2.79E-01	1.00E+00
descriptive statistical information of the TM							
art. mean	1.34E+05	1.35E+05	1.34E+05	1.34E+05	1.34E+05	1.73E+05	1.34E+05
std. dev.	3.14E+05	3.82E+05	2.58E+05	2.80E+05	1.75E+05	3.92E+05	2.46E+05
minimum	0.00E+00	0.00E+00	0.00E+00	0.00E+00	0.00E+00	0.00E+00	0.00E+00
maximum	1.37E+06	1.94E+06	1.09E+06	1.09E+06	6.29E+05	2.29E+06	9.99E+05
median	0.00E+00	0.00E+00	0.00E+00	0.00E+00	3.47E+04	0.00E+00	2.31E+03
mode	0.00E+00	0.00E+00	0.00E+00	0.00E+00	0.00E+00	0.00E+00	0.00E+00
samples	2.54E+02	2.54E+02	2.54E+02	2.54E+02	2.54E+02	1.98E+02	2.54E+02
skewness	3.30E+00	2.56E+00	1.84E+00	1.30E+00	6.50E-01	2.72E+00	9.78E-01
kurtosis	1.06E+01	5.81E+00	2.24E+00	2.23E-01	-1.25E+00	6.54E+00	-6.18E-01

5.2 Visual Interpretation of the Full TM Scene

For the map generation, the original unprocessed images were transformed into a real coordinate system; namely the UTM. For these transformations, reference points were taken from the 1: 50000 scale topographic maps as shown in the next chapter. Through this processing, versatile color composite maps were created.

Basically any digitally manipulated color image production is a color-additive/ color-composite in its nature and it does not include all the wavelengths of the sensed environment. The statistical manipulation in the digital image processing also changes the albedo characteristic values of the original image substantially. The original Landsat TM data from 1986 was system corrected prior to delivery. Due to that, after mosaiking the four sub-scenes together, there was an albedo and contrast difference between the scenes, which misguides both the visual and digital interpretation process. In the further analysis activity, this factor was taken into account, see [map 1].

In the case of data from the multi-spectral scanner on the Landsat TM series satellites, the conventional approach is to assign bands 1, 2 and 4 to the colors blue, green and red, respectively. In this way terrain with healthy vegetation, which has a high intensity of reflected radiation in the near-infrared (band 4) and very little reflection in the yellow-orange (band 1) will appear red. The IR color image is a composite of bands 2, 3, and 4 in blue, green, and red. Vegetation reflectance becomes progressively lower for bands 5 and 7, and cultivated fields are increasingly darker in these images. In the thermal IR image (band 6), despite its low spatial resolution, topographic details are recognizable. The dark signatures depict the relatively cool temperatures and are associated with shadowed, high inclination slopes which have mainly east facing aspect. Bright signatures manifest the relatively warm features and are associated with a direct sunlight hit, shows also high inclination slopes with mainly south facing aspect. In the analysis, contrast enhancement processing was applied to each band separately, which enabled to achieve a maximal contrast gain.

The full scene Landsat TM band-combination (4,7,5) in RGB was processed and transformed into IHS domain. The saturation stretched and back-transformed image of this IHS processed data resulted in a more contrast rich image as is shown in the [map 1]. In it there is a richer contrast difference between the pixels of the scene especially in the extreme lower and higher grey values.

The city Addis Ababa, which lies on the edge of the western escarpment of the main Ethiopian rift system, is distinctly shown with its bluish-iron color. On the Entoto mountain-chain around the northern, northwestern, and western area of Addis Ababa a narrow belt of green vegetation is observed which mainly is attributed to the eucalyptus tree concentration there.

In the escarpment, south and southeast of Addis Ababa, there is a lack of major faulting, and therefore, of morphological definition. The Menagesha mountain west of Addis Ababa and the Zikuala conic crater lake, far south to Addis Ababa, can be observed very clearly. Immediately southeast of Addis Ababa, on the way towards the rift valley, lies the Yerer mountain. South to the Yerer mountain, trachyte and trachytic tuff with some rather recent superimposed basalt cones extends into the rift valley region.

The rift floor is characterized with a light tone, which is attributed to the recent sedimentation process in the vicinity. In this dry-season image, the area is covered with dry brown grass. The area also includes a substantial area of post-harvest and fallow cornfield as well as meadow field.

In the central north part of this image, the Shenkora highland, there is a violent retrogressive erosion of the western escarpment of the central rift valley highland region. The area of erosion in the western and eastern escarpment is generally darker in the color tone than the rift valley floor where a recent fluvial deposition occurs.

The big agricultural fields of Wonji, and Metehara sugarcane farm and the upper Awash farm (in the far northeastern part of the image) can be distinctly observed due to their deep red color. They are all irrigation farms, which are based on the Awash river. Except for the above-mentioned plantation, there is no meaningful irrigation in the area and the whole farm activity of the rural population fully relies on the seasonal and timely rain.

Associated with the volcano Fantale, northeast of the image, the many hyper alkaline silicic lava are fairly shown in the image. South and west of Fantale, which is completely superimposed with the other surrounding material reflectance in the spectral image, the plain is dotted with numerous cupolas (ceiling of dome) or blisters, hollow domes composed of extremely scoriaceous and altered lava encrusted internally.

In the southeastern part, the big Dixsis-Farm, which is exclusively rain feed, is seen as a big rectangular surface due to its post-harvest distinctive reflection. The morphology of the whole scene is best understood if one uses this map in combination with the figures [Figure 3], [Figure 4], [Figure 5] and [Figure 6] in chapter 1.

5.3 Surface Water Availability

Differences in the appearance of surface water are readily apparent on multispectral images. A spectral characteristic of water is dependent on the wavelength and is a superimposed result of the molecular nature of the water, its depth, turbidity, and the presence and concentration of biological organisms such as algae in it. The concentration, size, shape, and reflective/refractive index of suspended particles determine turbidity and increase the amount of energy backscattered from the water body. Delineation of the location and spatial extent of surface water is most successfully carried out by using data acquired in the near-infrared spectral area.

The visible wavelength data (0.4 to 0.7 μm) had provided information on certain physical conditions within lakes, rivers and wetlands. Water has a very high transmittance (low absorption coefficient) for the visible spectrum. Near-infrared portion of the spectrum is almost completely absorbed by water and very little energy is back scattered which implies water bodies to have a significantly lower reflectance than terrestrial features throughout this portion of the spectrum. Because of this phenomenon, surface water appears black on reflective infrared images except for cases of extremely high near-surface turbidity, [Table 9]. The aerial extent of the surface water body is relatively easily and accurately delineated on remotely sensed data acquired in the near infrared bands. For Landsat MSS, the ranges 0.98 to 1.08 μm is the best for delineating surface-water area except its coarse spatial resolution. Using the band combinations (7,5,4), in RGB on-screen river vectorization was done.

Table 9. Qualitative estimate of relative turbidity of water from Landsat MSS images (after [Moore1978]).

relative turbidity	tone of image				hue of image
	band 4	band 5	band 6	band 7	
none	dark	dark	black	black	black
slight	medium	dark	black	black	dark blue
moderate	light	medium	dark	black	medium blue
heavy	light	light	medium	dark	light blue
very heavy	light	light	light	medium	white

For the surface water mapping and study, both the Landsat TM and MSS were first PC transformed and Inverse PC transformation was performed with the stretched principal components. This processing enabled to get best contrast especially in the extreme low and high ends of the spectrum as will be discussed latter on in detail. After such a processing, an unsupervised classification was done on the IPC data.

The above processing on IPC MSS image overlaid with vectorized drainage data has resulted the [map 2]. This map shows the surface water bodies and drainage features of the full Landsat MSS scene. The vectorized rivers were overlaid on the classified result in order to show the main river courses and their tributaries. In this map, the shadow of the cloud (mainly in the southeastern part) does also have blue color after the classification which eventually lead to a false conclusion about the location of a real surface water. The brown line is the watershed to the north with the Blue Nile basin, to the south and south east with the lakes region and the Wabi Shebele basin, (vectorized from the hydro geological map of Ethiopia, EIGS).

The Legedady and Gefersa dam on the western escarpment, are the two main drink water sources of the capital city. The Abasamuel dam, which serves mainly for a hydroelectric generation lies to the south of Addis Ababa. This is followed by the Koka dam, the biggest dam of the rift valley region used for both hydro electric generation and irrigation. The semi circular crater lakes around Debre Zeit are clearly shown in the map. The lake Zway in the southern part of the map, and the newly expanding lake Beseka (lake Metehara) in the far northeastern, are the two natural lakes located within the Ethiopian rift valley floor.

The Awash river which originates from the highland areas west of Addis Ababa, first flows straight into the rift valley but after reaching the watershed with the Lakes region, to the east of Koka dam, it changes its direction in the center of the main rift valley towards the east. After passing the Koka dam it flows up to Awash Melkassa, far northeastern corner of the map, along the eastern flank of the rift valley. The Kessem river starts at the Shenkora Highlands and flows along the foot of the western escarpment. The two rivers flow almost parallel to each other for approximately one hundred kilometers. The Kebena river in the far north merges with Awash and Kessem, in the Awash Melkassa area.

5.4 Landsat TM and MSS Scene Study area Cutout

For further detailed study, a cutout of the six topographic sheet coverage between approx. 8°30' and 9° Longitude and 38°40' and 39°30' Latitude was prepared. This includes a wide range of the western escarpment and the rift valley area. On this cutout, PC-analysis, band ratioing, and convolution filtering were performed. In the preparation of the noise corrupted original data, histogram equalization and linear stretching were applied. For the whole dynamic range, from 0 to 255, the DN values were computed once and saved in the lookup table so that these values were easily available for the further computation.

By manipulating the frequency distribution (histogram) of an image in such a way that the output image should have a horizontal characteristic line, different results were generated. These were optimized images for visualization and further digital classification. In this processing, in addition to the horizontal characteristic-line, which represents an equal pixel distribution throughout the whole dynamic range, gausian normal distribution approximation was also applied.

In the histogram equalization process, the histograms with minimal attendance frequency were lost. After several empirical experimentation and comparison with the ground-truth data, the most relevant signal intervals from the original histogram were extracted and summarized in [Table 10]. Due to this selection, for those images which did not have relevant values at minimum and maximum, the linear stretching method was preferred. Here, the mean value of the histogram was placed on the middle value of the whole dynamic range, for 8-bit data 127, and the input image was stretched on the whole displaying dynamic range.

By using the newly acquired Landsat TM+, attempts were made to perform change detection analysis. But the results were not satisfactory. The main cause for this may be the far smaller size of the farm plot on the ground and the coarser resolution of the TM pixels as a whole.

Table 10. *Grey value segmentation for the different images and bands used in this work.*

imaging satellite and band number		Relevant pixel values aggregation-interval		
	type of the band	cloud (noise)	relevant data (signal)	cloud (noise)
Landsat MSS	band 4	0-15	16-52	53-127
	band 5	0-13	14-78	79-127
	band 6	0-20	21-78	79-127
	band 7	0-25	26-83	84-127
Landsat TM	band 1	0-60	61-119	120-255
	band 2	0-18	19-68	69-255
	band 3	0-18	19-102	103-255
	band 4	0-19	20-109	110-255
	band 5	0-7	8-177	178-255
	band 6	0-120	121-199	200-255
	band 7	0-3	4-94	95-255
SPOT	panchromatic	0-18	19-50	51-255

5.4.1 Applying the Principal Component Analysis (PC)

A Principal Component, PC, transformation is based on either the scene covariance or correlation matrix, which produces new variables known as components or axes that are linear combinations of the original variables. Each component contains data uncorrelated with the other component.

The PC transformation was useful for better aggregation of the information from each of the bands. The Principal Component depicts and allows to extract more information than the original bands. For performing this PC transformation, the covariance matrix of the MSS and TM images were used, see tables [Table 7] and [Table 8]. The higher values in the upper and lower portion of the matrix clearly show the high correlation of the spectral bands.

The PC transformation is commonly applied to all available bands of the satellite data at one time. The resultant output images emphasize distinctions between surface materials, but because each new output image is a linear additive combination of all input channels, the individual images or their color composites often are not interpretable in terms of a sensed phenomenon, as is possible with ratio images. The method PC-transformation is best used as a tool for distinguishing between rather than identifying lithologies. However, an evaluation of the transformation values will allow identification of physical objects on the ground which will often be manifested in the new resultant image.

While performing the PC-transformation, the individual component-value will be ordered in such a way that, the first component will get the majority of the total scene variance, with latter components getting less and less of the total scene variance. Moreover, the typical high correlation of the spectral bands to each other, which are responsible for the less variations in shades of the colors on a standard TM and MSS false-color-composite image, are eliminated or at least subdued.

Table 11. Eigenvector matrix, eigenvalue and percent variance accounted for in each principal component of Landsat TM image bands. All bands were first recomputed to a common pixel size of 30 meters.

Landsat TM bands	PC1	PC2	PC3	PC4	PC5	PC6	PC7
1	0.25	0.13	-0.40	0.10	0.11	-0.16	-0.20
2	0.20	0.22	-0.33	0.04	0.04	-0.16	-0.65
3	0.43	0.19	-0.56	0.04	0.15	0.13	0.53
4	0.19	-0.52	0.09	0.50	0.29	-0.57	0.06
5	0.60	-0.34	0.30	-0.13	0.28	0.54	-0.18
6	0.12	0.32	0.24	0.16	0.01	0.01	-0.39
7	0.49	0.05	0.23	-0.51	-0.42	-0.50	0.11
eigenvalue	2206.69	343.21	325.25	158.14	76.91	45.12	20.54
percentage	69.09	10.75	10.18	4.95	2.41	1.41	0.64

The Principal Component 1, PC1, provides a measure of the overall albedo, illumination and brightness. The eigenvector of the first principal component PC1 in the above table contains a high loading from bands 5, 7, and 3 followed by the bands 1 and 2 and the least contribution is with the band 6 amounting 0.12, see [*Table 11*]. It accounts for nearly 69% of the total scene variance. The term loading could be understood as the proportion of each band pixel linearly combined to create the new pixel value for the respective principal component, [Colwell1983_1]. The substantial positive loading of PC1 is manifested in its ability to best reflect the geomorphologic structural setup of the study area including man made structures. It is a good index of the effects of topographic influence on illumination. Thus, better than any of the individual original bands, it is a valuable tool for terrain morphologic analysis. Very detailed and recent lineaments around lake Koka (NNE-SSW) and west of the lake Beseka can be easily observed.

Principal Component 2, PC2, contains the positive loading of bands 6, 2, 3, 1 and 7 contrasted against the negative loading of the bands 4 and 5 over a wide range of loading intensity. Basically this component contrast suggests that materials showing strong contrasts between these groups of bands will be emphasized in the resultant image as the materials that exhibit the strongest contrasts will show the greatest separation in grey shades on the image. It provides an image that is similar to the color composite and has the advantage of enhancing vegetation features. This component is also weighted from the water bodies of the study area. It accounts for 10.7% of the total scene variance. The PC2 shows that the plateau and escarpment surfaces, which contain dense stands of coniferous and deciduous vegetation, and the lowland cultivated areas stand out in dark shades. The rift valley bottom and river gorges, which contain sparse grass and sage bush, are shown in shades of light grey to white color.

Principal Component 3, PC3, contains a large positive loading of the TM band 6 against the strong negative loading of band 3 and also accounts for 10% of the total scene variance. With 0.089 and –0.091, the loading of band 4 and 7 are relatively insignificant. This component can be interpreted as defining materials that have a strong contrast between the negative loads from bands 3, 2, 1 and positive loads from bands 5, 6, 7. Examination of the generalized reflectance curve in [Figure 14] on page 32 shows that typical vegetation exhibit such contrast between these collective bands. Since the thermal band 6 is also positively added in this component, the cooler, shadowed, and wetted areas are much more represented than in the other components. Here farm plots are best shown including the towns and their satellite villages. With PC3, vegetation and water body of the area is better displayed.

Principal Component 4, PC4, contains negative loading of band 7 against a large positive loading from 4 and 6. It accounts for 4.95% of the total scene variance. The contribution of the other bands to this component is insignificant. The strong contrast of band 4 and 6 against 7 can mainly be attributed to vegetation with some contribution from the morphology and man made structures such as streets and small towns. Drainage structure and main roads are good retained. Local volcanic structures can also be clearly seen. The main regional structure and the underground controlling of the surface geomorphology are here apparent. In the Principal Component PC4, the background material such as soil and geology are better emphasized. PC4 enhances more the "true nature" of the chain of mountains, their structural setup (morphology), background material (soil) as well as the areas of erosion and sedimentation.

In the Principal Component 5, PC5, the bands 3, 4, and 5 are strongly contrasted against band 7. Here, the geomorphology and vegetation is strongly suppressed. River drainage basins and regional structures can be clearly delineated. This Principal component accounts only for 2.4% of the total scene variance. The fine background structure of the area is clearly represented in this principal component.

Principal Component 6, PC6, shows main positive loading from band 5 contrasted against bands 4 and 7. This PC also accounts strongly for the geological structure of the study area and the background soil and rock are effectively contrasted. The morphological and man made structures are subdued. This component accounts for 1.43% of the eigenvalue of the total scene but delivers us a vital information concerning the tectonically controlled structure of the area. It shows similar fine structural setup like PC5, but this one is with a more distinct structure of gorges and rivers.

Principal Component 7, PC7, accounting for 0.64% of the total scene variance, depicts nearly equal loading from band 3 contrasted against band 2. This should define hydro thermally altered rocks. But since the area is intensively used for agricultural purposes and is mainly covered by thick tropical soil, it is difficult to certify the existence of such rock types.

By taking out the effect of morphology (PC1) the combination of histogram equalized components (4,3,2) in RGB, [map 3] gives the best color-composite false color image of the area in relation to land use. The Principal Components 2, 3, and 4 were the most suited combinations for the land use detection and identification. This combination suppresses the topographic effect. The natural lakes were clearly shown with a varying color from light blue to deep dark blue depending mainly on their turbidity, shallowness and the algae living in them. The subsistence farm plots and meadow areas are detectable as a bright yellowish spot. The smallness of the farm plots prohibits the further mapping at this scale. The green vegetation mass, bush and natural forests, along the hill slopes and gorges in the image do have a light to deep red color. Generally, deep red hues indicate broad leaf and/or healthier vegetation while lighter reds signify grasslands or sparsely vegetated areas. Densely populated urban areas are shown in light blue similar to traditional infrared aerial photography.

In the color composite map of the histogram equalized Principal Component combination (3,1,2) in RGB, [map 4], the water body in the area is much more vividly shown and the morphology of the area is also best displayed. Areas with green vegetation are represented with the reddish lila color. The volcanic intrusion and mainly from NE to SW directed structures can be easily mapped. Man made structures such as towns and roads are also identifiable. The recent volcanic centers of the area namely Yerer, Zikuala the lava outflow west of Nazereth are distinct.

5.4.2 Inverse Principal Component Analysis

The inverse principal component is the mathematical back transformation of the principal component data into its original RGB domain utilizing the inverse of the principal component matrix [Soha1978]. This amounts to reapplying the principal component transformation on the normalized contrast stretched principal component variables using a row or column transpose of the eigenvector matrix. In doing this, the data value is re-rotated from the principal component axis back to the original band axis. The stretched inverse PC is best suited for visual interpretation because it gives back the detail of each grey level, especially for the extreme white and extreme black pixels an extra differentiation possibility. The green vegetation covered area is distinctly differentiable on this color composite map, (3,2,1) in RGB, as is shown in the [map 5]. From this result, the main lineaments were traced and prepared for further study. These lineaments are mainly attributed to the geologic structure of the area. This image was further processed using IHS in order to become better contrast in vegetation distribution, as will be discussed latter on.

The [map 5] together with the color composite maps of the next subchapters gave the best information for further structural analysis and rose diagram data extraction.

As can be seen from the above discussion, the advantage of inverted PC in this processing includes:

- maximizing the spectral separability of the sensed materials through the inverse principal component transformation because of the normalization applied to the principal component data,
- subduing the "noise" inherent in the later components of the PC transformation,
- making the resultant images more interpretable in a phenomenological context,
- the color composite images of the inverted principal component results are exaggerated or vivid than the original band combination and bright or dark color materials (bright or dark hue), because they appear white or black on the standard enhanced image respectively. In the inverted IPC, there is more color delineation due to color separation capability of this method.

5.4.3 Edge Enhancement With Smaller Convolution Matrix Size

Because a regular high-pass spatial filter removes most of the albedo information from an image, its use is not always advisable for object identification. Using a very small (3 by 3 or 5 by 5 matrix) smoothing filter on the first difference can help to suppress random noise and to study various trends together for better discrimination. Edge enhancement, with a small convolution kernel, is a technique that can be considered a first order correction to the modulation transfer function of an image. The kernel size that should be used for edge enhancement was selected based on the number of edges or how busy the particular image is by trial and error.

After a high pass filtering by using a 7x7 kernel matrix, the original image was added to the filtered image with 100% feedback. A resulted color composite (7,4,2) in RGB, retains the benefits of the infrared bands yet presents vegetation in familiar green tones. Band 7 helps to discriminate the moisture content in both vegetation and soil. Urban areas appear as light green. The olivine-green to bright-green hues normally indicates forested areas. In the color composite (7,4,2) all lineaments are not well contrasted. The result from this processing was not as good as expected. However, a screen shot of this processing was taken and used for fieldwork in the first part of this research.

5.4.4 Convolution Enhancement for Regional lineament Analysis

On the Landsat TM and MSS original data, a number of convolution operations with different convolution matrix size and type were performed. The spatial information in an image can be considered as being composed of low and high frequencies. Due to this, different information-extraction techniques were developed which enhances the high frequency and low frequency parts separately. The low frequency component is usually represented by large areas of constant brightness, albedo or color information. High frequency information consists of brightness changes over a short spatial dimension that occur because of contrast in slope attitude or topographic features or contrast in brightness at boundaries between geological units. This fact is exploited also in other image processing methods such as band ratioing as well, as will be shown in the next sub chapters.

After many trials, the three tow-dimensional spatial filtering kernels were found to be appropriate to generate best structural image maps. Namely the kernel sizes 31x31, 51x51 and 101x101. In the resulting image maps, both the regional and intermediate-level linear features were enhanced. The image from the 101 by 101 kernel-processing was less grainy than the others, and therefore lineaments are here more clearly defined. Part of the graininess seen in the maps, which were produced using the 31x31 and 51x51 convolution matrix, was due to the harder contrast stretch and because of their small dynamic range which was applied on the original image. The 51x51-convolution result shows the recent sedimentation area structure more clearly than the result from 101x101, which had simply smoothed it. For the area around Koka dam and in the main rift valley, the 31x31-convolution matrix gives a very good structural information.

By comparing the convoluted TM and MSS images, with the help of the above three convolution kernels, it is found that the result from the MSS image had more regional information than from the TM. This may be due to the availability of more detailed information on the Landsat TM, which undermines the more regional information in it, than the Landsat MSS which is more coarser in its resolution. Therefore, the MSS image was used for the lineament study in this work and the [map 6], [map 7] were generated.

The color composite in the [map 7] represents an updated inventory of those linear elements marked by morphological, tonal and textural variations that can be referred to major throws, scarp faults, faults and fractures. It is immediately noticeable that certain areas seem to be less affected by faulting and fracturing than others, and that some appear completely unaffected. These facts are related to the nature and age of the various terrain types of the region. More precisely, in the main rift valley region, in large areas blankets of recent incoherent alluvial and lacustrine deposits obliterate the underlying bedrock structure.

Even though the whole study area is highly populated and agriculturally used, man made structures could not be properly identified and mapped here on the processed color composites. This may be caused due to the combined effects of the dryness of the imaging month, the geometry and the small size of the most farmland pieces, often with 20x20 and less meters of size which are mainly cultivated with the subsistence farmer families.

Along the flanks of the rift especially in the north-central, northeastern and southwestern sectors of the region, the terrain is characterized by poorly organized drainage pattern strongly influenced by the structure and a high density of preferentially oriented lineaments. On the image, this portion of the study area is characterized by a relative smooth texture and a variability of grey tones, but the distinctive characters remain the faulting and fracturing along the NNE and NW directions.

5.4.5 Band Ratioing

As it can be seen in several publications, spectral band ratioing is a proven technique, which allows identification of geologic materials based on the recognition of diagnostic absorption bands. Band ratioing is useful because it minimizes the effects of topographic slope, aspect, and albedo differences between rocks and enhances the signal difference which arises in reflectivity between bands, which are diagnostic of various surfacial materials.

The ratio band3/band4 was chosen to distinguish the differences between rocks, which have relatively gently sloping spectral curves between the two bands and vegetation which do have a marked increase in reflectance between bands 3 and 4. Low ratio values (black to dark-grey shades for low ratio values) indicate vegetation, where as high ratio values indicate absence of vegetation.

Very often the band ratio band5/band1 was selected to define limonitic rocks, which have a marked falloff in reflectance in band 1 compared to band 5 because of ferric-iron absorption. It is to remember that vegetation is better captured in band 5 than in band 1. Due to this, the high-ratio values (light grey or white) will indicate the presence of limonite or vegetation, whereas low-ratio values may indicate their absence.

The ratio band5/band7 was chosen to distinguish materials displaying absorption features in the 2.2 μm region. Hydro thermally altered rocks containing argellic minerals have absorption bands in this region while unaltered rocks lack these features. Although vegetation displays considerably lower reflectance values compared to rock reflectance, their ratio values was similar to those of altered rocks. Thus high-ratio values indicate argellically altered rocks or vegetation. Both argellically altered rocks and vegetation will appear in light grey to white shades when processed in single band. The band5/band1 and band5/band7 ratios individually cannot be used to map limonite or argellic rocks because of the ambiguity related to vegetation.

Additionally, the following image enhancement methods (for vegetation index) were applied on the MSS image:

- Vegetation index: (VI) = (MSS7-MSS5)/(MSS7+MSS5)
- Soil brightness index (SBI) = 0.43*MSS4+0.63*MSS5+0.59*MSS6+0.26*MSS7
- Green Vegetation Index (GVI) = -0.29*MSS4-0.55*MSS5+0.6*MSS6+0.49*MSS7

The following neochannels were built from the Landsat TM data:

- NEOVEG = 128+128(TM4+TM3)/(TM4+TM3)
- VEG = 100 + 200(TM4+TM3)/(TM4+TM3)
- VEG*TM7 = 128 + 128(TM7-VEG)/(TM7+ VEG)

The [map 9] of the Landsat TM ratio combination enhances substantially those, in part generalized, geological and geomorphological formations of the area. Here the main erosion/deposition surfaces are clearly differentiated.

Some of the main regional lineaments were traced and overlaid on it for better emphasis, [map 10]. In the [map 10] of the MSS, band ratioing maximized the vegetation spectra and the morphological setup information while suppressing the background material reflectance of the area. It best discriminates the vegetative areas from less vegetative or dry areas. In this color composite the greenness (green mass) in some locations can be attributed to:

- enough underground water which comes from the northern higher elevations and respective local maxim,
- inaccessibility of the forest areas by the people and as a result all year availability of the natural forest,
- its unsuitably for grazing and farming activities or
- spectral misplacement due to the shadow created by the morphology.

5.4.6 Surfacial Tectonic Feature-Map and Rose Diagram of the Lineaments

Some of the major faults, that were previously mapped on the geological map of Nazereth, are identifiable on all of the above results. Some of the indicators that seem to imply this on the newly processed maps above includes straight drainage channel segments, straight contacts between erosional and depositional features, straight valleys in hard rock areas, and distinct hue changes on opposite sides of linear features in both morphology and adjoining surfacial materials. However, the above discussed image processing result show a much more lineament and other geomorphological information than it is shown on the geological and hydro-geological maps from the EIGS, [map 11] and [map 12].

In a common interpretation process, the interpretation procedure of a linear structure is to plot lineaments as dotted lines on the interpretation map. Field checking and reference to existing maps helped to identify some lineaments as faults. The new linear features on these maps may represent:

1) previously unrecognized faults,

2) zones of fracturing or

3) lineaments unrelated to tectonic structure such as lithological contact zone.

An overall look at the distribution of the structural element reveals three characteristic general trends. In the northern portion, moving from W to E, the trend is generally NNW oriented which gradually changes in the central region to NW direction up to E-W and ENE-direction. In the southern portion - especially in the south central zone - the general direction is NNE. Fairly long faults with major throws mark the internal zones of the structural systems emphasizing the axial zones of the rifts. They mark a zone of traversal-weakness.

The above produced maps, mainly the maps [map6] to [map10], but also [map4] and [map5] had provided a substantial information on the structural framework of the rift valley with particular emphasis on the possible relations between locations and alignments of the volcanic centers, the directions of the major tectonic features and the geometry of the rift's margins.

These lineaments were vectorized on the screen in order to find out the lineament pattern and distribution. The values of the lineaments were then saved as vector files. By using a program module written in the programming language C and the vector files saved above, each value of the lineament including its length, angle (by taking an angle interval of 10°) and direction was computed and saved separately for further processing, see appendix 3. These were finally used as an input to the MATHLAB[iv] "rose" function for drawing the rose diagram.

[iv] MATHLAB is from MathWorks, Inc. Software.

The function rose plots an angle histogram, which is a polar plot, showing the distribution of:

- internal friction angle,
- stress[v] intensity,
- count of lineament, and
- length of the lineaments in 36 groups of bins in the range [0, Pi].

A total of over 3000 lineation were drawn from the [map9] and [map10]. The loss of lineation, especially in the southern and the central part of the area, is mainly due to surfacial deposits, which mask the bedrock exposure and in some places due to the very recent lava flows.

The distribution of the lineation occurring in the whole study area is shown in the rose diagram, [Figure 18].

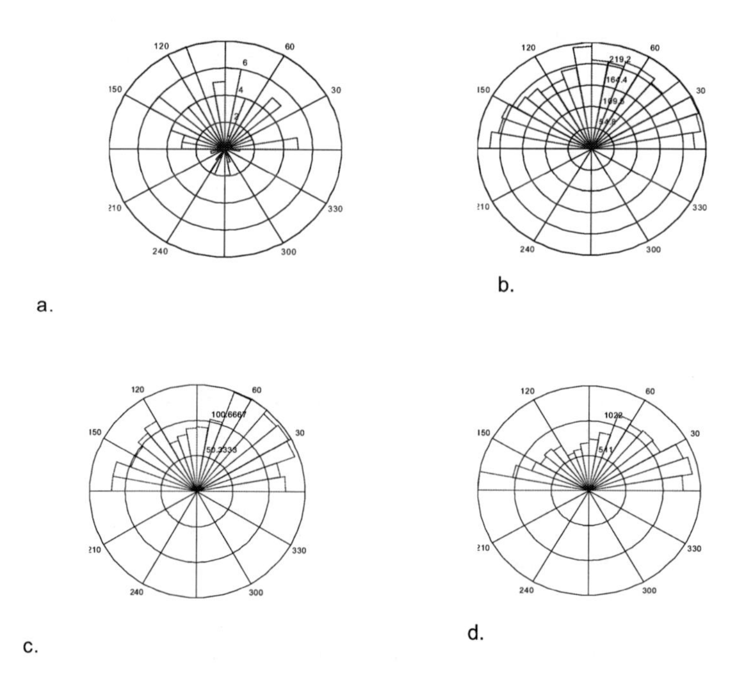

Figure 18. Rose diagram showing a) lineament angle of internal friction b) stress intensity c) lineament number count and d) lineament length respectively, based on the Landsat MSS image.

[v] Stress is generally any geological force which acts to deform rocks. The three major types of stress are Compression - a stress that acts to press or squeeze rocks together, Tension - a stress that acts to stretch a rock, or pull a rock apart, and Shear - a stress which acts tangential to a plane through a body, causing two contiguous part to slide past each other. The main cause of the stress in the study area is the triplate divergent motion (a tension), but can also be the volcanic activity (with its root cause in the earth mantel), and/or isostasy.

The main angle of internal friction for this area is 120° followed by 90° and 150°, see [Figure 18a].

The stress intensity diagram on [Figure 18b] shows the main stress direction in the ranges up to 30° and around 90°. This diagram shows that the stress intensity is relatively high in the other directions too.

The lineament occurrence count (frequency) approx. 30° and 60°, as is shown in the [Figure 18c] predominates the other directions, followed by 170° and 125°. The longest lineaments, which often extend over tens of Kilometers, are predominant in the 170°-180° interval followed by 10°-30°, see [Figure 18d].

The comparison of the azimuth frequency distribution and lengths of the four sectors shows a strict correlation with the dominant peak. The NE-SW system appears clearly to be the preferential lineation trend present in the above figure with the sole exception of the NW-SE sector, where a slight azimuth shift can be observed. The N30E system, appears clearly to be the preferential lineation trend followed by N60E, N150W and N120W respectively. This trend, spatially well distributed in all the various rock formations, can be related to a recent regional stress. It is also frequently coinciding with the alignments of the recent fissural volcanic centers.

The lineament frequency, length and direction inherits the potential capacity of underground water circulation.

5.4.7 Lithological Mapping Using Classification

In the preceding discussions, different textural filters have been applied on the Landsat images (original bands and neochannels) in order to create resultant images containing enhanced geological information, structures and textural signatures.

A classification of Landsat TM image, which was processed by a 101x101 convolution matrix was done in order to determine the lithology and lithological boundary (rock distribution) and its mapping.

For [map8] the TM images were first convoluted with 101x101 matrix moving average kernel in order to eliminate the high frequency component of the original data. Using it as an input an unsupervised classification with 50 classes was performed. In these 50 classes it was possible to detect even the different plots of the subsistence farms which may be of different crop type, post-harvest land, fallow area, meadow, small chick-pea farms etc.. This classification result were then visually analyzed and recomputed into 6 different classes in order to concentrate only on the main geologic-geomorphologic representative features. To achieve the same goal, an unsupervised classification, with a similar procedure as discussed above, was applied on the Landsat TM ratio (x2/x5, x1/x2, x1/x7) in RGB, [map13].

From the two classification results, it is hardly possible to differentiate between the lithologies and map any meaningful boundary. For a better comparison see the cut outs from the original geological map which is presented in the [map11]. This absence of any meaningful lithological differentiation and boundary can mainly be attributed to the high thickness of the upper soil, vegetation cover and intensive agricultural activity in the area.

Comparison of the hydro geological map of the Nazereth area [map12] with the unsupervised classification result, [map8] shows some conformity for the highland areas. These are marked on the hydro geological map as impermeable layer of N2r (the alkaline and per alkaline rhyolite domes and flows from early Pliocene), and Qwra (alkaline and per alkaline rhyolite, trachyte, domes and flows from Pleistocene). The rest of the area was marked as permeable mass on the hydro geological map and the classified image includes structures such as volcanic centers, horst and graben formations in the later area. According to this geological map, the most frequently occurring Alaji basalts of the early Cretaceous/Miocene – which have a thickness of 1000 meters or more - are shown to be moderately permeable.

A thorough study of the maps [map7] to [map13] allows the conclusion that the available hand dug wells on the SW of the hydro-geological map may have to get their water recharge substantially from the escarpment NE of it. This should be further investigated using special applied geophysical methods such as Very Low Frequency (VLF), electrical and electromagnetic methods and hydro geologic tracing methods.

6 Geographic Information System Database Information Management

In the past two chapters we had seen the different methods, which are applied for image processing. There, different thematic maps were generated. In this chapter, condensed literature study will be presented about the Geographic Information System, GIS, and the dependency of information on scale. The ARC/INFO[vi] technology for GIS implementation will be discussed. At the end, the necessary steps for creating a GIS database with their repetitive steps and the implementation of Arc View[vii] for its management are presented as a flow chart diagram.

The Object Oriented Data Modeling, OODM, concept which was extensively used in the programming languages such as Smalltalk (starting as early as 1983) and Java, from Sun Microsystems, is now finding its way more and more into GIS implementation. Unlike the traditional spatial database engines - which store spatial and attribute data in proprietary, and OpenGis[viii] data formats and in relational database management systems - the object oriented design more closely mirrors the real world. The object oriented approach shows the advantage of stronger and more flexible modeling, data integrity and higher security.

In a typical GIS implementation, under utilizing the available software at the department, six processes are integrated together, namely relating information from different sources, data capture, data integration, projection and registration, data structuring and data modeling. In general, it uses proprietary storage mechanisms for the geographical data. GIS can be defined as an organized collection of computer hardware, software, geographic data, and qualified professionals designed to efficiently capture, store, update, manipulate, analyze, and display all forms of geographically referenced information, [ESRI1994_4].

With a GIS, the capabilities of traditional database query will be extended to include the ability to analyze data based on their location. The overlay feature of GIS and the ability to compare different entities based on their common geographic occurrence is the key feature of GIS. Like conventional Database Management Systems, DMS, the geographic analysis system is seen to have a two-way interaction with the database. Thus, while it may access data from the database, it may equally contribute the results of that analysis as a new addition to the database. For example, we might look for the joint occurrence of lands on steep slopes with erodable soils under agriculture and call the result a map of "soil erosion risk". This risk map may be derived from the existing data and a set of specified relationships. Thus the analytical capabilities of the geographic analysis and the Database Management System, DBMS, play a vital role in extending the database through the addition of knowledge of relationships between features, see [ESRI_3], [ESRI_5].

The spatial data analysis in a GIS is the mechanism by which the user interacts with the GIS, and contains the functions that allow spatial reasoning to take place on the data in the GIS. Simple spatial query functions such as points in a polygon and differential distance calculations are fundamental to GIS technology, and the compilation of complex spatial

---

[vi] ARC/INFO is the product of ESRI Inc.

[vii] ArcView is product of ESRI Inc.

viii Open interfaces and protocols defined by OpenGIS® specifications support interoperable solutions that "geo-enable" the web, wireless and location-based services, and mainstream IT, and empower technology developers to make complex spatial information and services accessible and useful with all kinds of applications.

queries incorporating information from separate layers or coverage of GIS data is one of the important goals of spatial data systems. The analysis systems also allows for the creation of new GIS information (layers) by performing reasoning (such as aggregation over an area) on existing information of the systems. GIS analysis provides the ability to easily determine point in polygon membership, to compute relative distances, or to perform different operations on different layers of data. The GIS technology provides the engine that handles the graphic display and tabular output of the results of analysis. With the help of GIS, Graphic User Interfaces, GUI design and implementation become relatively easy. Complex GIS systems such as ARC/INFO have modules included as their integral part macro languages for further automation, see [ESRI_6].

The data input system is the input mechanism by which existing analog maps or new data is digitized into GIS and converted into the appropriate GIS format. There are many approaches to this task, and because both the digitizing and data conversion take a large amount of time and effort, it is potentially the most labor intensive of all GIS related tasks.

A data management system is composed of the mechanisms by which data is stored, retrieved and manipulated by the computer. Typically there are two parts to the data management system, though in some systems the same engine can be used for both parts. The first part of the data management system is the mechanism for dealing with the spatial information of the GIS data. The second part of the data management system is the mechanism for dealing with the tabular attribute data that is associated with each piece of spatial information [ESRI1994_2].

The ARC/INFO GIS system used in the present analysis, uses this two part data management mechanism, the spatial data is handled by ARC sub-program, and tabular data [attribute data] is handled by INFO sub-program which can be programmed to interact with other external DBMS such ORACLE[ix], SYSBASE[x], Microsoft SQL[xi], etc. This characteristic of the ARC/INFO program permits a reliable integration of traditional database management systems possible. At present, there are actually considerable amounts of attributary data administered with different databases across the country which could easily be integrated to handle large amounts of information. The ARC/INFO GIS provides versatile tools for collecting, saving, extracting and processing diverse data of geographical, geometrical or attribute nature. Its object orientedness and relational nature enables to model highly complex natural and social processes in our environment.

---

ix ORACLE Database is product of the Oracle Corporation.

x SYBASE is product of the Sybase Inc.

xi Microsoft SQL is product of the Microsoft Corporation.

6.1 Cumulative Impacts on the Environment and its Quantification

Cumulative impact can be understood as the impact on the environment which results from the incremental impact of the action when added to other past, present, and reasonably foreseeable future actions regardless of what agency or person undertakes such other actions. Cumulative impacts can result from individually minor but collectively significant actions taking place over a period of time. The area's cumulative impact on the environment, in general, is a superimposition of our activity, locally and globally, and the dynamic nature on its own. The incremental effect of these minor human activities will accumulate over a period of time. As a result, two critical issues will emerge, namely, the spatial scale and temporal scale of the cumulative impact and conceptualization of proper analysis method for its mitigation. In a longer term prospect, the evaluation and study of proposed projects and resource utilization within the context of entire drainage systems could be managed more meaningfully and in its integrity.

Traditionally, impact analysis widely relied on the use of analog overlay maps and often considers only a singled out location and the short-term consequences of land management activities usually several years at most and only a subset of a drainage. However, the issue of spatial and temporal effects do have a wide range of appropriate scales to consider that depend upon the question asked. For example, a domestic water user in a village might be quite concerned about changes in turbidity during a single storm. Changes that require a decade or more to become evident are long-term processes, for example, filling of pools with sediment or reduction in the supply of large woody debris that, in turn, causes the streambed to become unstable. The cause of the recurrent drought in the countries different regions is also such a long-term process. It is observed to repeat itself almost every 10 to 15 years. Such a reality is one of the manifestation form of a cumulative impact.

While implementing GIS, to use overlay maps, it is necessary to prepare maps that show the position, nature and extent of natural and human attributes of an area. Attributes, which have to be mapped for such research, include surface water bodies, agricultural land, wetlands, settlements, and cultural resources. The features mapped are those which are expected to be sensitive to the study project. Using this direct and collateral impacts can be identified, broadly and generally, by superimposing a map showing the proposed development and associated projects onto the composite map. Overlay mappings are useful for identifying impacts and comparing alternatives for all types of development and investment activities that finally manifest themselves as a cumulative impact.

The GIS technology may help greatly in the management and analysis of large volumes of data, allowing better understanding of environmental processes and better management of human activities to maintain economic vitality and environmental viability and mitigation of negative cumulative impacts on the environment.

6.2 Information as a Scale Dependent Quantity in a GIS Analysis

Scale may be defined as the mathematical relationship between the size of objects as represented on maps, or on images or other remotely sensed data, and the actual size of the objects themselves. An optimum map scale is determined by the specific problem setting. Selection of the appropriate scale is influenced by the resolution of the data, and both scale and resolution are application dependent. The scales at which data are collected and analyzed directly influence the level and kinds of information that may be obtained. Information derived from data collected at a particular scale is dependent on that scale. [McCarthy1956] strongly emphasizes the scale-dependent nature of information as follows: "... conclusions derived from studies made at one scale should not be expected to apply to problems whose data are expressed at other scales. Every change in scale will bring about a statement of a new problem, and there is no basis for assuming that associations existing at one scale will exist at another".

According to [Stone1972] a determination of the number and limits of the scale classes is the key step of the multiple-scale approach. The method by which the scale classes are selected will depend on the objective and complete field observations, followed by a careful analytical testing of various scales in comparison with the data available from all other sources, and the selection of the smallest or/highest scales wherein faithful generalizations may be made toward the initial objective or the study. The thrust of this approach is to develop a methodical procedure that guarantees consideration of all scales. One of the principal uses of the small-scale imagery is to locate potential areas of interest to be examined in greater detail by progressively larger scale remotely sensed data and/or other means. Small-scale data may often be meaningfully integrated to the data obtained from progressively larger scale remotely sensed data or from ground surveys of the area.

When dealing with information obtained at different scales, it is important to consider the degree of aggregation represented by each scale. Broadly stated, small-scale information requires an aggregation of data while large-scale information requires sub-division. Data aggregation occurs in the temporal, categorical and the spatial domain. The same consequences occur regardless of the domain in which the aggregation occurs. It is vital always to remember that data aggregation levels must be appropriate for the phenomenon being studied. It is unnecessary and costly to use data that are more specific than the level of analysis required. Even though the available data may have higher resolution, there may be relationships or classes contained within the data that might require that the data be "smoothed" or aggregated in order to detect trends that could be lost in highly divided data. When too many data is aggregated into a single resolution cell, relationships may begin to obliterate the expected result, resulting in artifacts. Results based upon these artificially created relationships can misrepresent the actual realities on the ground.

6.3 The Spatial Data Concept in ARC/INFO GIS Implementation

The ARC/INFO data model is a composite of different data model concepts and principles. ARC/INFO employs a hybrid data model called the geo-relational model to represent geographic features. Location (spatial) data is stored using a vector or a raster data structure. Corresponding descriptive (attribute) data for each geographic feature is stored in a set of tables. The spatial and descriptive data are linked so that both sets of information are always available. Geographic features are arranged and categorized to provide organization in the database, see [ESRI_1], [ESRI_2].

In ARC/INFO a continuous surface can be represented using a triangulated irregular network (TIN), GRID[xii], or a lattice. GRID is the frequent method for representing raster data. GRID can represent areas, points and lines, as well as continuous surfaces. On the other hand the coverage is the main method for managing vector data. It is best suited for accurately depicting the location and shape of points, lines and areas. The coverage stores line and polygon features topologically. This optimizes data storage by reducing coordinate redundancy, and facilitates a number of key spatial operations such as polygon overlay. Coverage, GRID and TIN surfaces are geo-referenced using the same coordinate system and map projection, see [ESRI_3], [ESRI_8].

Geographic data sets are the data models supported by ARC/INFO to represent geographic information. Each geographic data set is characterized by the features it depicts, its method for representing shape and location, and its utility for various geographic operations. Each model, namely coverage, GRID, lattice, and TIN are structured and saved as a subdirectory folder. These folders are partially created and managed by the ARC/INFO program. In ARC/INFO Software there are different tools available, with which each individual model can be selected and processed.

Making spatial data usable requires correcting the coordinate data to make it free of errors and topologically correct. This was accomplished by establishing the existing spatial relationships (constructing topology), identifying errors, correcting errors, and then reconstructing the topology. The root mean error for the processing in this work was set under 0.00002 mm. The on-screen digitization of a coverage in this study was done using the ARCEDIT module and the flow diagram [Figure 19] shown in the next page, see [ESRI_7].

[xii] GRID, TIN; ARCEDIT, Lattice are modules of the ARC/INFO software.

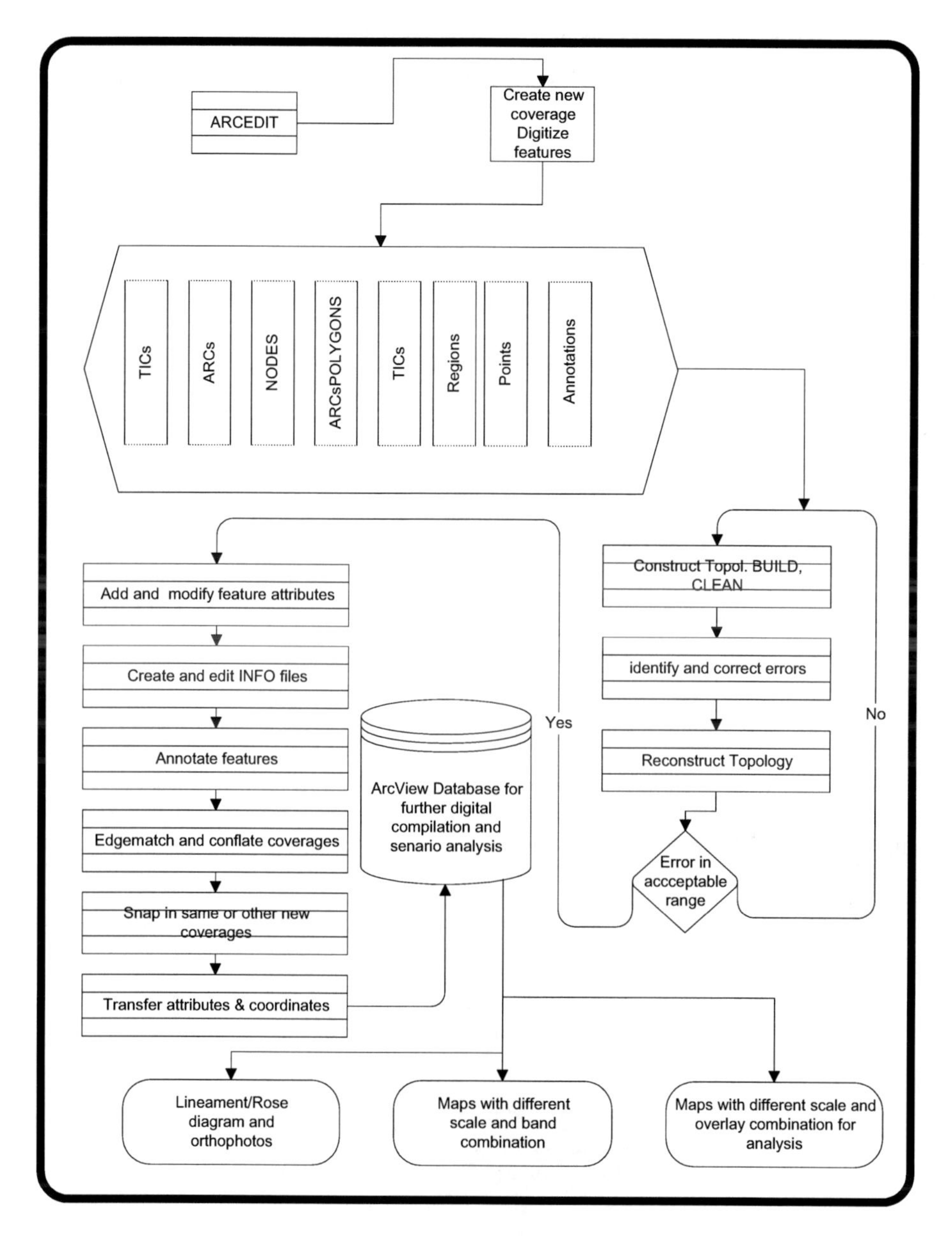

Figure 19. Editing a coverage using the module ARCEDIT of the ARC/INFO software.

6.3.1 Linking Attributes to Features – and the Georeferencing

On maps, symbols and alphanumeric characters convey descriptive information. Often, the textural information provides a way of accessing additional information organized in other files. The map then becomes a powerful tool for referencing information. The same concept applies to the spatial data models. One powerful capability of GIS lies in its ability to build a link between the spatial data and the descriptive data, (which are considered as attributes of the spatial one). GIS is used to maintain the connection and restless integration between the spatial features and their descriptive data [ESRI1994_9].

The link between the spatial definition of features and their corresponding attribute records is created using a unique identifier of a feature, which associates the attributes with the feature coordinates, maintaining a one-to-one correspondence between the spatial records and the attribute records. Once this connection is established, it is possible to display attribute information, or create a map based on the attributes, stored in the attribute table.

The relational concept can be applied to more than just keeping track of features and their attributes. Any two tables that share a common attribute can be related. In digital maps, spatial relationships are depicted using topology. Topology is a mathematical procedure for explicitly defining spatial relationships. Topology expresses different types of spatial relationships as lists of features (e.g., as area is defined by the arcs comprising its border). This allows processing of larger data sets and substantially faster processing, [ESRI_1].

6.3.2 TIC Points and Georeferencing in GIS Processing

For any database to be useful for spatial analysis, all parts of the database must be registered to a common coordinate system. Most maps display coordinate data by conforming to a recognized global coordinate system. Map projection ensures a known relationship between locations on a map and their true locations on the surface of the earth. Although map projection does not constitute a geometric error, it does require a geometric transformation of the input data, and this can be accomplished by the same operations and compensate for distortion in the data.

Transverse cylindrical projections such as the transverse mercator use meridians as their tangential contact, or lines parallel to meridians as lines of secancy. Their lines of tangency then run north and south, along which the scale is true. For the Universal Transverse Mercator (UTM) system, the globe is divided into sixty zones, each spanning six degrees of longitude. Each one has its own central meridian. In Ethiopia the transverse merkator projection is uniformly used and it lies on the 37-north zone, [ESRI1994_9].

A set of TIC points with a common coordinate system were established which enabled the rectification of the whole data to the common georeferenced coordinate system as will be discussed in the next chapter. The following flow diagram in [Figure 20] was used for the generation of the TIC points.

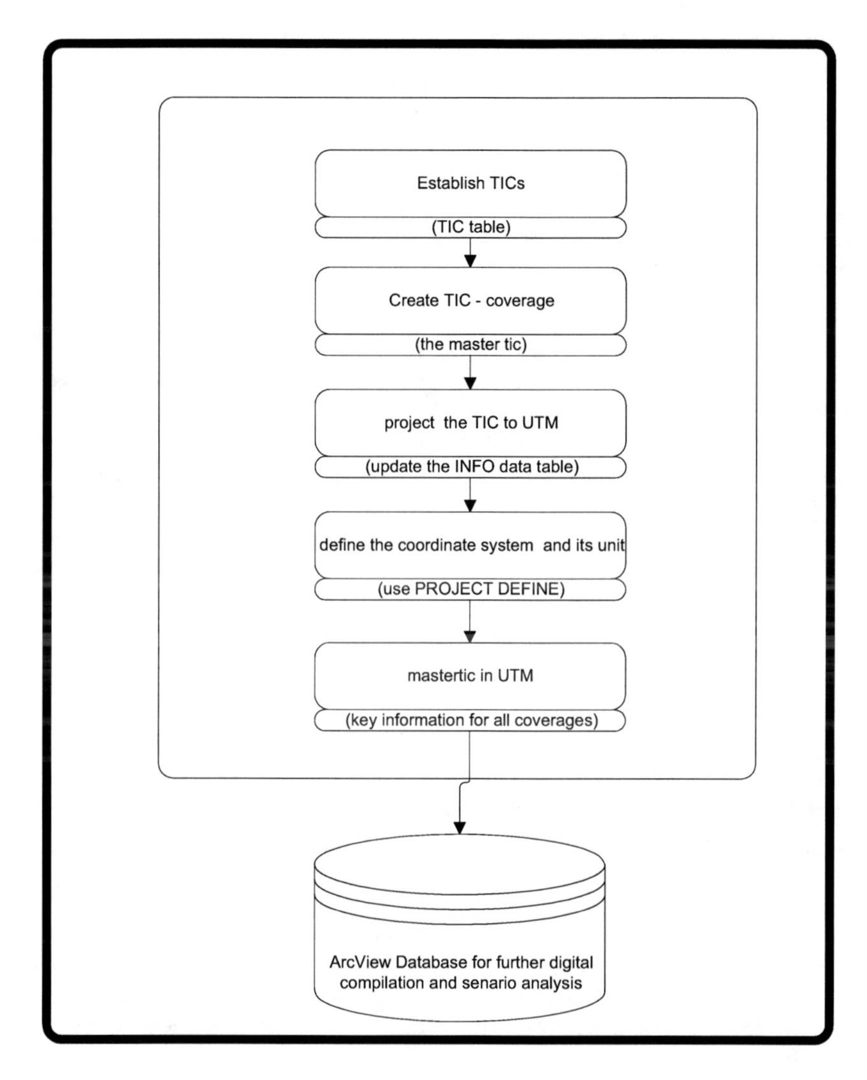

Figure 20. Creating Master-TIC table and determining its real world coordinate location using the ARC/INFO software.

6.3.3 Digital Versus Tablett Digitizing

The digital onscreen vectorizing and digitizing have shown a lot of advantages against the digitizing using a digitizing tablet. Among these merits are:

- the accuracy of the digitizing process was better than that from the digitizer tablet because of constant fixation in the whole digitizing process once it is scanned and saved,
- the continuation of digitizing can be done without any problem after taking an off brake and one is not restricted to the room with the digitizer since it can be done on any computer with the appropriate software facilities,
- due to the zoom in / zoom out functionality of the digitizing programs it is possible to digitize with an adaptive tolerances than the primary fixing of weed tolerance snapping tolerance, grain tolerance etc. This increases consistency and work quality which otherwise is not achievable,
- the time which is necessary for correction and problem shooting is generally much less than for the digitizing tablet,
- an onscreen digitizing is up to 80% faster than on the digitizer tablet and
- the scanned data will allow, at least for the road (red color), water body (blue color), to use an automatic color separation enabling an automatic vectorization.

Among the demerits of the on-screen digitizing the beginning expenses for the scanning, the necessity for high RAM-Memory, enough hard disk availability and faster processor are the outspoken ones.

6.3.4 The TIN Model Value Generation

TIN is one of two data structures in ARC/INFO best suited for representing the shape of a continuous surface, the terrain. TIN is an alternative to the raster data model for representing continuous surfaces. It allows surface models to be generated efficiently to analyze and display terrain and other types of surfaces. The TIN model represents a surface as a series of linked triangles. Further detailed explanation is given by [Chuchip1997].

In a typical GIS database building, after successfully creating the coverage, the next step will be to build the TIN model. The triangles are made from three points, which can occur at any location. This contrasts with the raster where points are spaced evenly in the lattice or GRID models. The TIN model creates a network of triangles by storing the topological relationships of the triangles. The fundamental building block of the TIN data model is the node. Nodes are connected to their nearest neighbors by edges, according to a set of rules. Left-right topology is associated with the edges to identify adjacent triangles, [ESRI1994_10].

In ARC/INFO, the data format of the ARCEDIT module is not the same with the TIN module. Therefore a small program, in C-programming language, was written to convert the ARCEDIT data format into the TIN data format, see appendix 3. Using this conversion program, the Z-values of the TIN module were generated from their ID-number values of the ARCEDIT. In doing this, a new table with (x, y, z)- fields were created for the three dimensional TIN-modeling and the further study. This TIN modeling further enabled the generation of the data for slope and aspect processing as will be discussed latter in the next chapter.

6.3.5 Layer Automation in ARC/INFO and the Workspace Concept

The main method of ARC/INFO to physically organize a geographic database on a computer is by using a workspace.

All of the GIS work was performed in a local ARC/INFO workspace with a complete read and write access. Path names are used to display predefined geographic data location in the computer data system so that performing logical and spatial queries on the GIS database is secured. Workspaces can usefully organize layer automation. A common method is to create one workspace for each layer to be automated. In that workspace, a copy of the layer is made at each automation step so that the state of coverage at a particular stage can be recovered at any point. Naming conventions were used to indicate the step in the automation process. This method allows each layer to be automated and tracked in its own workspace. Steps often need to be repeated, and this method provides an efficient means to do so. The flow chart in [Figure 21] shows the workflow used for building the GIS database in this study. In the next chapter, the processing of the data including detailed discussion will follow.

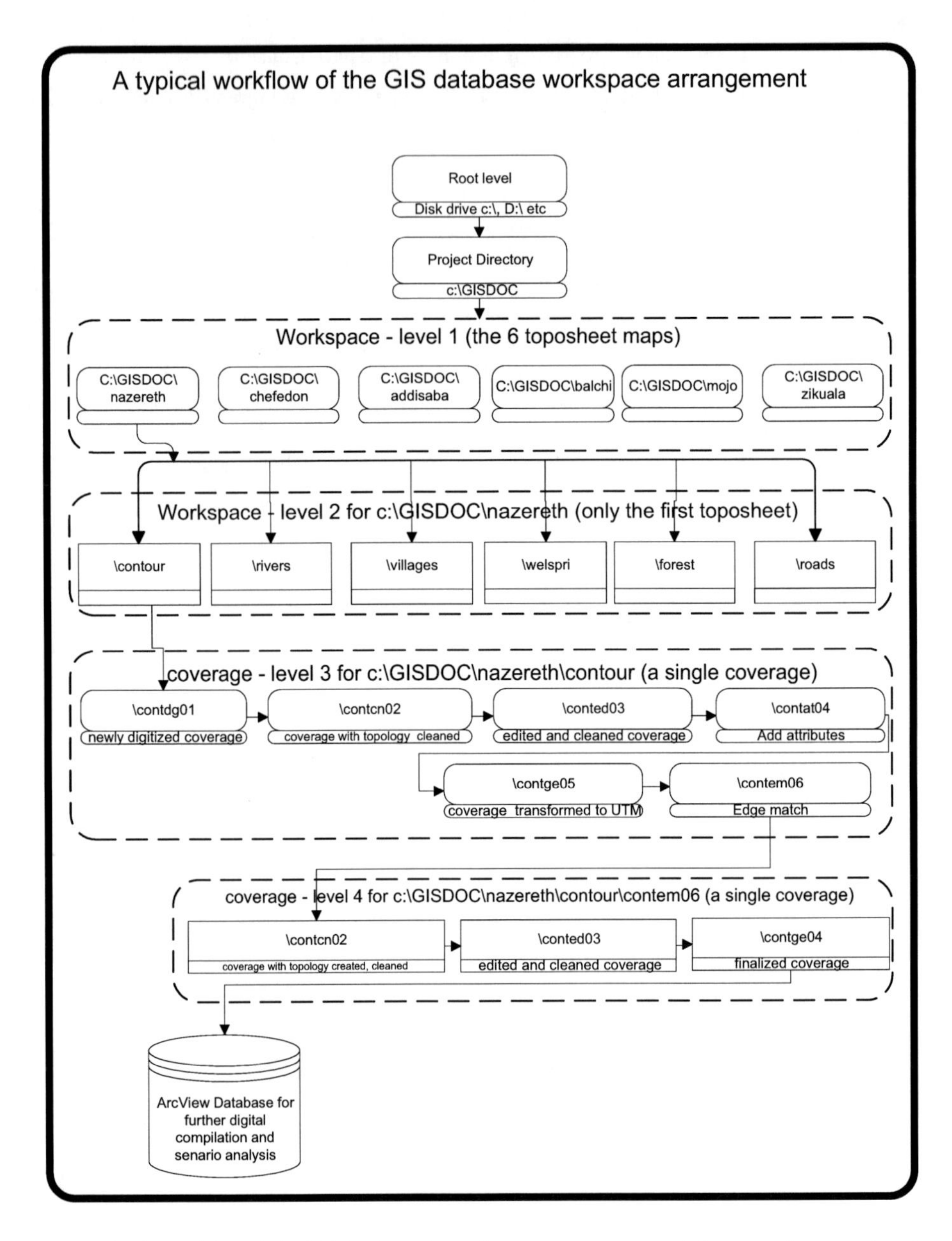

Figure 21. The workflow of the GIS database, its workspace arrangement and the processing procedure for this study.

7 GIS Database Building Interpretation and Discussion

In the last chapter, the basic concept of GIS in relation with the intended application was discussed. A flow chart for the processing was presented. In this chapter the scanned six-top sheets were digitized. For this, different fault tolerances and digitizing schemes were applied. A GIS database was built. Different coverage, GRID, and TIN models were created. Geographical coverages such as elevation, slope, aspect, landform and slope position were created and discussed. Image maps created in the image processing were integrated. A digital spatial database was designed and developed, spatial analysis was performed, different maps, were created and a report was generated. Combining of several layers of data was done to study relationships between them.

7.1 Digitizing and Vectorizing the Topographic Maps of the Study Area

The first step in using the available topographic map was to scan them with the optimal resolution/data size ratio. To decide this, three different scanning qualities were selected and tested with the following parameters:

case 1) 300 dpi with 8-bit,

case 2) 400 dpi with 8-bit and

case 3) 600 dpi with 8- bit grey level values.

After many trial and errors, case 2) was found to be optimal for most of the work and was used for this study. The whole topographic sheets were then scanned with the above specification. [*Table 11*] below shows the scanned topographic sheets with their respective original TIF file size. Since it was scanned at once, possible geometric distortion and complication, which could arise from partial scanning, was avoided. These TIF files were then used for the extraction of the contour line, road, village, wells and springs, vegetation, lithological boundaries, and water shade coverage.

The data is then vectorized on the screen and further enhanced using ARC/INFO version 7.2.1 on an Intel Pentium III machine - Windows NT operating system.

Table 12. Scanned topographic, hydrogeology and geology map sheets with 400 dpi and 8-bit resolution.

Nr.	location	TIF-file size in MB
1	SE Addis Ababa 1: 50 000 scale	124
2	Zikuala 1: 50 000 scale	111
3	Chefedonsa 1: 50 000 scale	98
4	Mojo 1: 50 000 scale	105
5	Balchi 1: 50 000 scale	88
6	Nazereth 1: 50 000 scale	114
7	geology of Nazereth area 1.250 000 scale	260
8	hydrogeology of Nazereth 1:250000 scale	280
9	Hydrogeology of Ethiopia 1:2 mio. scale	300

At some places in the locations with higher gradient change in the z-axis, i.e. rapid elevation change, the contour lines are near to-each other and on the digitized data instead of having a line it tends to be a surface of mixed color. The lines were "melt" into each other. For such cases, it was necessary to take a higher resolution (800 dpi or higher). This in turn resulted in a TIF-file of a considerably big size, which was difficult to save on a single CD, and further processing was also more difficult due to the higher demand on the Read Access Memory (RAM) capacity.

In order to reduce the high bulk of the digitizing work, the color recognition and separation facilities of the ARC/INFO and Adobe PhotoShop[xiii] software were used. The RAM storage problem was partially solved using the color separation methods to distinguish the features depending on their color coding on the original map. An automatic color separation algorithm was used to separate the different geographic features from the scanned topographic sheet. In this processing there was a difficulty of distinctiveness especially in the contour lines. This was mainly caused by its brown color, which is mainly a mixture of basic colors. As a result, the contour maps were vectorized from the original scanned map without any color separation technology.

For the roads, rivers, and springs/wells the color separation brought a good result. The resulting data was then further enhanced using different color manipulation methods and vectorized easily.

Even though the villages were black in color in the original map, after digitizing, the color of some villages did show additional color mixture, which had complicated the automated separation of their color as black. Therefore, the villages were first color enforced and then separated from the scanned base map.

[xiii] Adobe Photoshop is the product of Adobe Inc.

After an on-screen digitizing and automated vectorization process, the final cleaning was done with ARCEDIT and different themes were created for each topographic sheet. Since the tolerances (0.00002mm) are set during digitizing it was not necessary to change here once more, because the maximum tolerance was set to the quite complex geometric feature on the topographic map while vectorizing. Any decrease in tolerance would cause a loss of these information.

Using the data handling steps discussed in the last chapter on [Figure 21], data dictionary (digital manuscripts) were prepared for the items depicted in [Table 13] below. The data dictionary maintains the names of attributes and a description of the attribute values and will be used as a reference in the consecutive processing steps. These digital manuscripts were then processed to create the coverage with their respective coordinates and attributes. At this point, each map manuscript had been digitized, and vectorized.

The accuracy of the digital representation of spatial data is mainly governed by the user requirements, the inherent characteristics of the source document and the instruments used to create it – which were available at the department. Positional and spectral accuracy cannot exceed those of the original data source whether it is a large-scale map sheet that has been digitized or data collected from an orbital remote sensor.

Table 13. The main data layers created and maintained in the GIS database.

geographic feature/ layer	**structure/feature type**	**spatial objects/ feature class**	**attributes**
geological	raster georelational grids	geological mapping of the area, cells	geological formations, lithological boundaries
hydro geological	raster georelational grids	hydro geological map of the area, cells	hydro geological formations
tectonic	coverage-vector arc node topological georelational	teconical structure of the area, arcs, nodes, annotation, tics	tectonical lineaments
drainage of the area	vector arc-node topological georelational	the river drainage of the study area, arcs, nodes, annotation, tics	stream class, rate of flow
settlements of the study area	coverage, topological gorelational,	label points, annotation, tics	name, size of the villages
geomorphology of the study area, terrain surface representation	TIN, surface, triangulated irregular network, aspect, shade index	x, y, z- nodes edges triangles,	relief of the study area
surface modeling and 3D-display, visibility, profiling	TINS and Lattices	x, y, z- nodes edges triangles (for TIN), x, y, z- points (for lattices)	slope, aspect, drape
image interpretation tectonic	raster, pixels bands, georelational	raster images as a map, orthomaps, images as attributes, change detection in the study area	different spectrally derived maps

7.2 Creating a Master TIC File for the Topographic Map Coverage

Establishing a master TIC file with master map extent is the most vital activity in georeferencing process. TIC points register coverage coordinates to a common coordinate system, and therefore, relate locations of features in a coverage to locations on the earth's surface. In the study area the geographic database map spans over six map sheets of 1:50 000 scale measuring around 28 x 28 kilometers each. To create a master TIC file, first the tics were numbered with a unique ID-number and their coordinates were determined based on the coordinate locations marked on the original map.

As discussed in the last chapter, the TIC registration will allow layer coordinates to be registered to a common real world coordinate system, in this case the UTM, relates the location of a feature in a layer to its coordinate value on the earth's surface. For the purpose of converting any coverage, GRID, TIN or image into a real world coordinate system, a set of TIC points have to be established. The first step in ensuring the coordinates registration is to obtain valid TIC coordinate locations. These TIC locations contains the ID-Number, x-, and y-coordinate values for each of them. Using them it was possible to transform the coordinates of layers from digitizer/pixel to UTM units. Although the TIC-files will not be used till the digitized coverage is transformed, it was important to identify the TIC locations before the coverage has been created, i.e. quite at the beginning. Location references for each TIC will be marked on the original scanned map using UTM coordinates. In this data processing, around 20 points were chosen for each digitized 1:50 000 map which allows to select best fitting coordinates from them.

For the creation of the master TIC, few points were chosen from each topographic sheet with a common overlapping corner points as is shown in the [Figure 22].

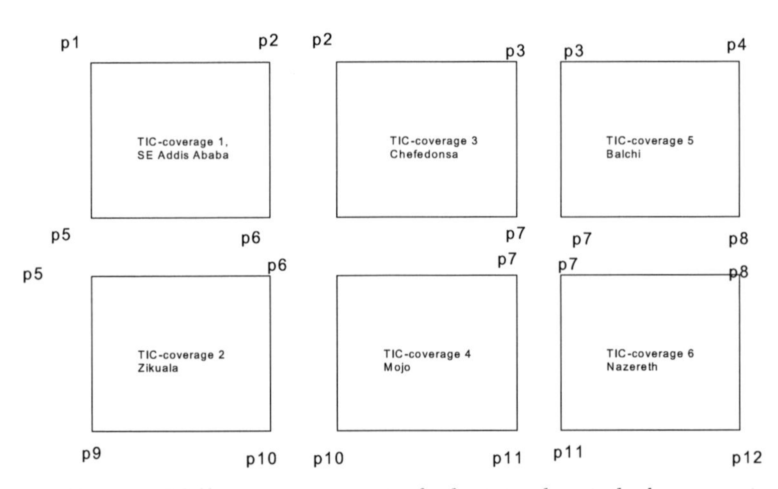

Figure 22. Master TIC file generation convention for the six top sheets in the data automation process.

In the above figure, the original topographic sheets from the Ethiopian mapping agency are placed in their successive order. On each topographic sheet, the adjacent locations of the TIC-registration do the same UTM value. The master TIC files with the master map extent are then established as is shown in table, [Table 14].

Table 14. Table of master TIC values for the study area.

	real-world coordinate TIC values (UTM mercator transformation)	
Tic_ID	**real world location- x value (in meters)**	**real old location y-value (in meters)**
P1	472530	994776
P2	499991	994765
P3	527466	994760
P4	554960	994768
P5	472530	967144
P6	499991	967142
P7	527434	994760
P8	554975	967162
P9	472493	939408
P10	499892	939493
P11	527434	967173
P12	555009	939529

After the TICs were properly registered, they were used in every georeferencing process from pixel value to real coordinate system. The master-TIC ensures that each GIS overlay is in the same coordinate system and matches correctly. TIC points are registration or geographic control points for each overlay allowing all its features to be registered to a common geographic control adjacent to a common coordinate system. Additionally, it ensures layers of the same area to be spatially registered using the same map projection.

7.3 Elevation and the Climate Variation

Based on the [Table 14] above, layers were created and saved in a common database. In the following subtopics, maps of different types for the elevation, slope, aspect, village and forest distribution as well as the drainage pattern of the study area were generated and results are discussed briefly.

The TIN processing delivers a three-dimensional representation of the topography. Topography exerts a stronger influence on the life condition of the people on the ground and resolution values on the remotely sensed data than any of the natural surface variables except for the areas of very flat terrain. The TIN-modeling further delivers us the input data for slope and aspect study.

As a matter of fact, the change in elevation causes a change in the local microclimate and plant/animal population, but not necessarily along the 20 meter increments which is available for this GIS analysis. The process of separating elevation along 200 meters was found to be more appropriate for this study. In doing so specific elevation data may have been lost through this generalization, but trends in the climatic/biodiversity distribution is observed to be strongly related to the gradual elevation change.

In the [map17], the elevation increases as the colors change from deep green to yellow, orange over brown to black grey. The minimum elevation is 1100 meters and the maximum 3100 meters above sea level. Both [map17] and [map18] show an elevation decrease progressively from NW- towards SE. There is a semi-parallel stepwise decrease in elevation in NE-SW direction and almost perpendicular to it in NNW-SSE direction. This trend is obliterated by several patches of younger magmatic intrusions throughout the SW and SE. In the shaded index [map20], the above facts are more clearly depicted.

As shown on the [map19], the climate zone of the area can be studied according to the elevation difference classes:

i. Kola (warm) climatic zone which covers the elevation up to 1500 meters, which lies strictly within the main rift valley (colored in deep-green and green including part of the light green) area,

ii. Woina Dega (temperate) climatic zone covers from immediate Kola to Dega (1800 meters to 2500 meters a.s.l.). In the study area, except the volcanic centers Yerer, Zikuala and Chefedonsa the rest lies in this climate. Accordingly, the locations Addis Ababa and Debre Zeit can be classified into this climate Zone. Traditionally this Woina Dega is expected to have sufficient rain and water, and

iii. except some peaks shown on the map, there is no Dega climate in this area.

For better representation, the transitional areas from Kola to Woina Dega (elevation 1500 – 1700 meters) were classified separately with light green color.

By applying the Kola, Dega and Woina Dega climate zones, we can see that well over 65% of the study area is in the Woina Dega and the rest in the Kola zone. Here, there is practically no Dega climate. By considering the elevation surface extension, the climates of Addis Ababa and Debre Zeit are representative for Woina Dega and Nazereth for Kola climate.

The main NE-SW directed lineaments and large parts of the western escarpment lies in the Woina Dega climate.

As discussed in chapter 3, the climate in Addis Ababa and Debre Zeit shows a substantial difference revealing the classification of Woina Dega as a vague-conglomerate of an environmental entity. Only the monthly mean minimum temperature of Addis Ababa and Debre Zeit shows similarity. This climate zone classification method does not allow further higher scale study. Within the Woina Dega climate, there is a substantial rainfall and temperature variation.

7.4 Streams, Rivers and Distribution of Wells and Springs

In the overlay map of elevation surface with the village and drainage vector coverage, [map21], the seasonal swamp areas are also included. Streams, channels, and other linear water features are represented in it. As discussed before, the drainage pattern of the area is mainly governed by the rift valley formation and its morphological structure. The study area has a locally variable drainage pattern which is partly caused by the sequences of irregular patches of volcanic cones and big lava outpouring centers interwoven with recent alluvial/pluvial sediments. The main eruption centers of the past volcanic activities are now the active denudation areas.

The major surface catchments of the area are mainly situated on the northern part with a predominantly dentritic, circular, and linear drainage forms reflecting the quite variable geology and the governing tectonic structure of the respective location. The overwhelming drainages are seasonal streams of the two main rainy season. The absence of any meaningful and systematic small scale dam for the rural population have left them to be fully dependent on a timely seasonal rain.

The [map22] shows the distribution of springs and wells. Their distribution concentration is seen to be mainly controlled by the NE-SW directed tectonic structure. This observation increases the expectation of under ground water flow through the main faults along the above main geological axes. Further more, investigation using hydro geological methods may reveal crucial information.

In the [map16], the distribution of the wells and the springs was overlaid on the Landsat TM ratio. Its combination with the village distribution depicts the location of wells and springs with respect to the villages. As shown on the village distribution on [map16] and [map21], there is an almost uniform density of population-distribution in all climate zones. It is vital to notice that there is virtually no water regulating or dam structure as a whole on the escarpment. There is also no protected reservoir area for drinking water.

7.5 Dwelling Pattern of the Study Area and Socio-Economic Conditions

The villages in the countryside are almost evenly distributed throughout the study area without any resource distribution consideration or other plan base. By considering the population distribution on [map19], there is a high intensity of population both in the Woina Dega (temperate) and Kola (warm) zone. More intense villagization is observed in the surroundings of the cities and smaller towns. In the central northern and eastern parts, the village pattern tends to show a parallel NE-SW alignment apparently controlled by the horst and graben formation. The [map23] which is a result of an overlay of villages over the aspect, shows their distribution to be remarkably concentrated on the horst formation. This situation is more clearer on the central and NE part. In the northern part of the study area, they tend to follow parallel pattern of the geomorphology with regular circular appearance surrounding volcanic centers. This considerable alignment of the villages with the existing horst formations especially along the main NE-SW fault direction shows the potential difficulty in modern surface water management and supply mechanism. The drainage intensity is higher on the highland areas than in the rift valleys showing the great problem of accelerated erosion due to the intensive subsistence agricultural activity of the people living in the highland areas.

Close investigation on [map16], [map21] and [map23] shows that the majority of the villages are located far away from big water locations. Due to this distance, the transportation of drink water takes a considerable amount of working time. Rain independent irrigation activities are unthinkable in such locations. This depicts the high vulnerability of those villages for weather fluctuation and misplacement of the seasonal rain. Taking the elevation as a factor for rain distribution, almost all of the population in this area is living under critical and chronical shortage of drink water, not mentioning the difficulty of rain independent agricultural production.

Even in Addis Ababa and Debre Zeit, which lies in the temperate climate zone, the pan evaporation is higher than the precipitation. There is virtually no protected water catchment area. Depending on the population distribution and by using the information given in the above mentioned maps, some optimization on environmental and surface water management can be done.

7.6 Natural Forest and Green Area

Acacias, eucalyptus, sycamores are the main trees that prosper on this climate regime. In this work it was impossible to classify the different species of forest due to the inadequate data resolution.

The natural forest is mainly distributed in very few locations which are rough and not easily accessible areas, [map13], [map15]. On [map13] The green color represents the natural forest covered area. This is a classification result on a first ratioed and vegetation spectrum maximized data. The [map21] shows the natural vegetation in red color. From the maps [map13], [map14] and [map15] no planned forestation can be seen among the villages. There is no systematic reforestation. Meaningful reforestation of the natural forest requires a long-term financial incentive to the local people and a proper planning.

As shown in the above maps and in [map18], the forest coverage is very modest. This is mainly caused by the high demand for fire-wood and for construction across the area. The necessity for more farm and grazing land is also the other cause for diminishing the forest. After comparing [map24] with the village distribution pattern in [map16], the following can be deduced:

- the overlay of forest distribution on the slope distribution shows that only a part of the less agriculturally used locations are covered with forest, which directly implies a potential rapid environmental degradation. The forest distribution overlaid on drainage pattern also shows a small overlapping of drainage surrounding with natural vegetation gallery,
- the unplanned use of the land had resulted in the dwindling of the natural forest and other natural resources. Even the high slope areas are used for agriculture which may ignite an erosion of the top soil causing a constant negative impact on the environment, and
- still by far the main source of household energy and light is wood and rural land use planning is completely absent.

In this work, the type of forest in the remote locations could not be easily identified. Additional data from higher resolution systems such as ICONOS and IRIS may bring better resolution for the identification and mapping of the forest types. In this regard, the unpublished works of [Lex2000], [Schneider2001] may also give supporting information on the possibility of mapping forest using remote sensing in Ethiopia.

7.7 Slope

The elevation layer is a quantity from which many landform descriptions were derived. The slope is one of the most fundamental quantity of landscape characteristics. It is a driving variable in many ecological studies and is directly related to soil surface depth, soil texture, internal soil drainage, soil group and depth to least permeable soil horizon. Slope steepness is included as a major factor in determining the soil erosion rate from a site. Slope stability is directly related to slope steepness [Sidle1985]. Slope length and angle affects soil depth, texture, profile and erosion of soil [Wishmeier1965]. Similar to the effects of elevation, flora and fauna species are often segregated along slope breaks.

The first vertical derivative of elevation is slope steepness [Evans1972]. Slope reflects the angle of inclination/declination of a land surface from the assumed horizontal line. The accuracy of each method of slope calculation varies depending on the quality and scale of the digitized contour maps, computed slope values, as can be expected, shows greater standard deviations in the high and variable morphologic relief than areas with the low one [Chang1991]. Slope steepness, measured as rise over run, was easily calculated using the ARC/INFO.

Once the slope data are computed in digital form, numerous transformations were performed on them to seek a variety of meaningful conclusions. The terrain of the study area shows a very diverse and wide range of slope steepness.

The slope data was classified along 10° increment with the ARC/INFO program. The resulting map for the study area is shown in [map25]. Light grey colors indicate the flattest areas approx. 95% of the total area (with an angle less than 10°); whereas blue areas represent the steepest sites (30°-40°). For better understanding of the slope-class distribution for the study area a slope value with 3° interval is shown in table, [Table 15]. Over 44% of the surface is flat (value –9999). The slope interval between 1° and 12° is 46.27% and above 12° amounts to 9.69% of the total surface. [Figure 23] shows the slope distribution with respect to its occurrence and respective sum of the covered surface area.

Table 15. The slope distribution and its respective covered area. Slope classes were used in 3° intervals.

slope interval in 3° rate	slope class	frequency count in %	area covered in %
(-9999. flat area)	0	not applicable	44.01
greater than 0 and less than 3	1	32.06	25.47
greater than 3 and less than 6	2	7.55	12.93
greater than 6 and less than 9	3	6.38	5.14
greater than 9 and less than 12	4	5.25	2.73
greater than 12 and less than 15	5	4.50	1.70
greater than 15 and less than 18	6	4.36	1.26
greater than 18 and less than 21	7	4.22	1.07
greater than 21 and less than 24	8	4.16	0.87
greater than 24 and less than 27	9	4.14	0.70
greater than 27 and less than 30	10	4.06	0.59
greater than 30 and less than 33	11	3.78	0.50
greater than 33 and less than 36	12	3.29	0.39
greater than 36 and less than 39	13	3.08	0.35
greater than 39 and less than 42	14	2.50	0.28
greater than 42 and less than 45	15	1.94	0.23
greater than 45 and less than 48	16	1.74	0.18
greater than 48 and less than 51	17	1.16	0.14
greater than 51 and less than 54	18	1.10	0.12
greater than 54 and less than 57	19	0.78	0.09
greater than 57 and less than 60	20	0.61	0.08
greater than 60 and less than 63	21	0.47	0.08
greater than 63 and less than 66	22	0.40	0.07
greater than 66 and less than 69	23	0.30	0.08
greater than 69 and less than 72	24	0.29	0.10
greater than 72 and less than 75	25	0.29	0.11
greater than 75 and less than 78	26	0.32	0.14
greater than 78 and less than 81	27	0.37	0.15
greater than 81 and less than 84	28	0.37	0.16
greater than 84 and less than 87	29	0.25	0.12
greater than 87 and less than 90	30	0.28	0.15

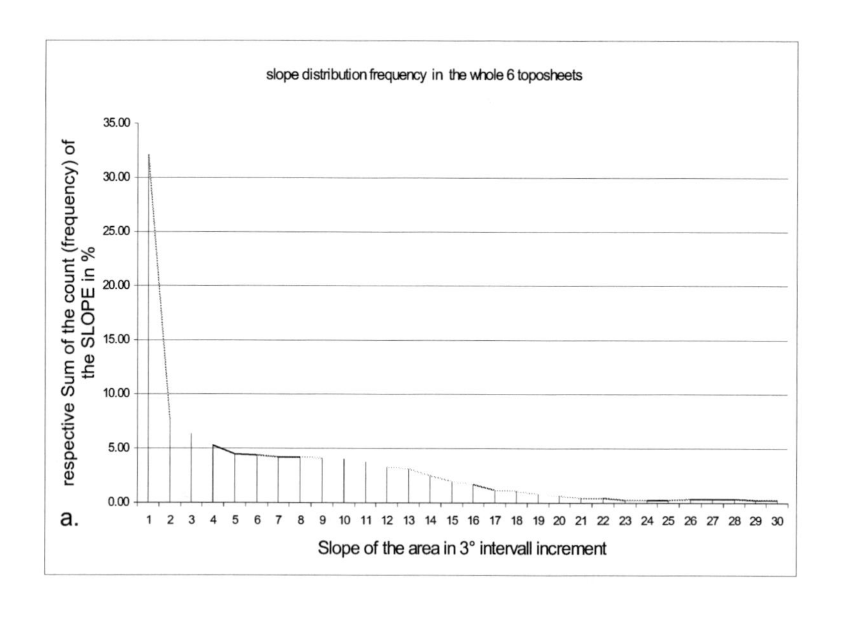

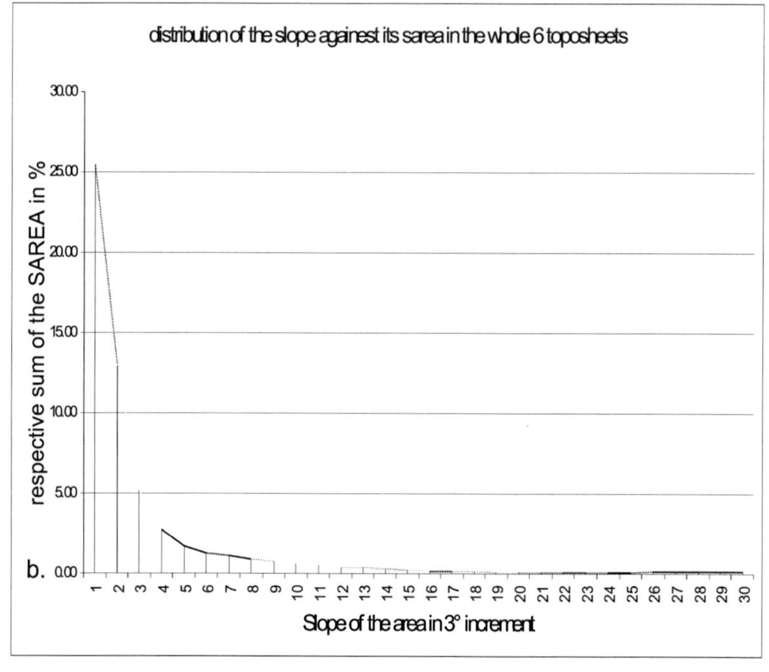

Figure 23. Slope of the study area with a slope increment of 3° a) with respect to the occurrence frequency and b) with respect to the respective covered total surface area, sarea.

7.8 Aspect

The first horizontal derivative of elevation is aspect [Evans1972]. Aspect can be understood as the direction a slope "faces" or the compass direction downhill from a point. Aspect is a circular function with 0° to the north, 180° to the south. It is the compass-direction to which the slope faces. Digital aspect computation may also depend on the scale of the digital elevation model and the relief of the terrain. As the raster size increases and/or the terrain becomes more flat, errors associated with aspect calculation will increase [Isaacson1990].

Many differences in landscapes can be attributed to differences in aspect. It is possible to have two or more locations placed at the same elevation, slope steepness and geological formation, yet the flora assemblage may be different if the aspect is different. On the windward side of mountains there is often much more rain fall than on the leeward side.

In [Table 16] and [Figure 24], the aspect was subdivided into 8-compass directions, excluding the flat value –9999 which accounts for 50% of the total study area. The south east aspect is the most frequent and covers the widest area. Then follows the north west and south compass directions. The smallest aspect compass direction is the north. In each of the aspect-interval, all the slopes between 0° and 90° are present, so as there is no dominant slope in a specific aspect interval. The distribution of the village and forest is not even among the aspect classes. The area which were represented in each aspect are almost proportional to the frequency of the aspect classes.

In [Map23] and [Map26] aspect of the area representing four primary compass directions is shown. The deep red shade in the latter map represents east, violet west, light green south, and light brown north directions, respectively.

Table 16. Distribution of aspect among the whole study area

Class Nr.	Aspect class number 45°		Frequency count	% of sarea	Min. slope	Max. slope
1	North	337.5 - 22.5	8.11	7.22	0.41	89.35
2	NE	22.5 - 67.5	8.48	7.02	0.36	83.92
3	East	67.5 - 112.5	12.93	13.42	0.25	89.36
4	SE	112.5 - 157.5	20.28	20.15	0.25	85.92
5	South	157.5 - 202.5	13.05	16.26	0.39	89.65
6	SW	202.5 - 247.5	11.01	11.99	0.43	84.25
7	West	247.5 - 292.5	11.94	12.22	0.4	89.53
8	NW	292.5 - 337.5	14.21	11.73	0.34	85.62

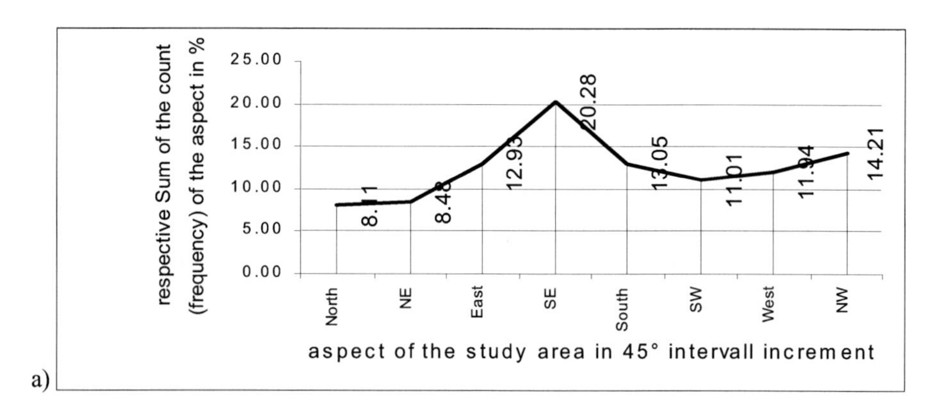

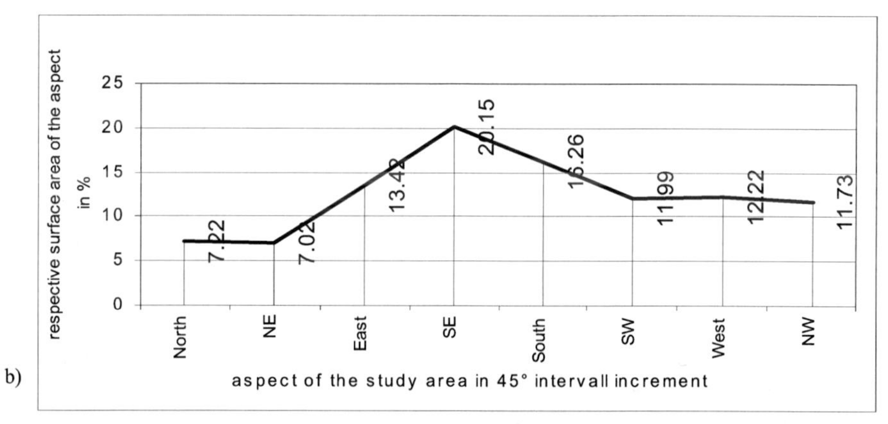

Figure 24. Distribution of a) the sum of the coount (frequency) of the aspect and b) the surface area covered by the aspect against the compass direction in 45° interval, after reducing the value 9999, which is 50.7% of the total surface area.

8 A Methodical Approach to the Introduction of GIS to Ethiopia

In the past chapters the application of remote sensing and GIS technology as a combined tool for the surface and ground water study in particular and an environmental study, in the broader sense, had been discussed. The next difficult task is the introduction of this technology to Ethiopia under the condition of scarce financial resource and trained man power.

In this chapter, a condensed literature review is done on the recent development challenges of the country and on ways of GIS implementation strategy. At the end of the chapter a proposal will be made on the possibility of price effective and optimal introduction of GIS to Ethiopia. The multi-participant approach is found to be best fitting as will be elaborated below.

8.1 Administrative Institutions and Potential Users of GIS in Ethiopia

In Ethiopia there are various governmental institutions, private agencies and Non Governmental Organizations (NGOs), which may potentially use this opportunity. The recent political power and administrative mechanism in Ethiopia is realized by the federal and regional governments at the higher level, woreda and kebele administrative units at the lower level of the hierarchy. This inherits different management and responsibility levels in any decision-making, planning, realization and quality control processes and procedures.

In October 2001, the Ethiopian government had announced new additional ministries that may help bring more efficiency and better work coordination. Among the new ones are ministries of rural development, capacity building, infrastructure, federal affairs and revenue with each of them having legal jurisdiction to coordinate the activities of related public offices. For instance the ministry of education, science and technology commission, civil service commission, Ethiopian management institute, Ethiopian civil service college and justice and legal research institute will be accountable to the ministry of capacity building, see //www.ethiopianreporter.com/eng_newspaper/Htm/No267/r267new2.htm.

The large scale introduction and implementation of GIS to the country may be better facilitated, due to the existence of these new conglomerate federal offices.

8.2 The Ethiopian Government Recent Sustainable Development Program and the Potential Role of Geographic Information System in its Realization

In the recently revised working document of over 200 pages[xiv], the government has clearly stated its intention and framework plan for the fiscal years 2002/2003 through 2004/2005.

The major and broad thrust of the strategy during the program's execution period are stated as[xv]:

- overriding and intentional focus on agriculture as the sector is the source of livelihood for 85% of the population where the bulk of the poor live. The government gives overriding primacy to the welfare of rural populance. Agriculture is also believed to be a potential source to generate primary surplus to fuel the growth of other sectors of the economy (industry),

[xiv] Ethiopia: Sustainable Development and Poverty Reduction Program. Federal Democratic Republic of Ethiopia. Ministry of Finance and Economic Development; July - 2002; Addis Ababa Ethiopia.

[xv] ibid. page i.

- strengthening private sector growth and development especially in industry as means of achieving off-farm employment and output growth (including investment in necessary infrastructure),
- rapid export growth through production of high value agricultural products and increased support to export oriented manufacturing sectors particularly intensified processing of high quality skins/leather and textile garment,
- undertake major investment in education and strengthen the ongoing effort on capacity building to overcome critical constraints to implementation of development programs,
- deepen and strengthen the decentralization process to shift decision-making closer to the grass-root population, to improve responsiveness and service delivery,
- improvements in governance to move forward in the transformation of society, improve empowerment of the poor and set framework/provide enabling environment for private sector growth and development,
- agricultural research, water harvesting and small scale irrigation and
- focus on increased water resource utilization to ensure food security.

The paper states further "... the strategy is built on four pillars (building blocks). These are: agricultural development led industrialization, justice system and civil service reform, decentralization and empowerment, and capacity building in public and private sectors."[xvi]

The development program and all these "building blocks" may need an extended availability of distributed information base.

The agriculture accounts for 45% of the gross domestic product of the country for the year 2000/2001[xvii]. By taking in to account the fact that only an estimated 3% of the countries food crop production is based on irrigation[xviii] and 85% of the population is a full time engaged private-subsistence farmer, it is not hard to see how heavily the countries rural economy is based on a seasonal timely rain. As it is observed in the recent past, the primary challenge to the subsistence farmers of the country, consequently the macro-economy, is the irregularity or total absence of the rain in the expected season – after each household have invested thousands of working hours on their farming fields, sawing their expensive seed and fertilizer - on which they can never have a direct influence. When the rain completely fails, the other wing of their economy - the cattle breeding – will soon come under a strong pressure due to the lack of grazing grass and water.

The working document shows further that, the government has planned to develop around 53 000 hectare of irrigation[xix]. Otherwise, the bulk of the strategy is inherently heavily based on the unknown and unforeseeable factor, namely a seasonal, timely and enough rain. Unless an economically feasible and sustainable introduction of water conservation/management mechanism for all climate regimes – and virtually for each groups of villages - is in place, it is hardly possible to attain the anticipated food security by the government.

[xvi] ibid. page iii.

[xvii] ibid. p. 33, figure 4.1.

[xviii] ibid. p. 87.

[xix] ibid. page 46.

In this regard it will be imperative to develop and implement a sort of nation wide drought-reversal and famine-preventive adaptive policy, which in its best scenario can be based on the introduction and routine implementation of "the state-of-the-art environment/water management technology coupled with a hydro/wind/solar electric generation" by using financial incentives and other business encouraging mechanisms for those highly specialized and well qualified private "will be" companies - by facilitating and allowing them to conclude agreements over a long period of time (possibly over several decades), making the work legally secured, profitable, attractive, and wholeheartedly implementing the fundamental property right - governed by the rule of law.

The implementation and realization of the well articulated three years framework plan of the government will certainly depend on how well and "reality based" the work on the ground will be done, the degree of transparency and accountability of the institutions involved in it, combined with the emplacement of a highly professional and efficient quality-control mechanism at the coupling interfaces or nodes.

Here, a reliable and robust information infrastructure may have the most crucial role. There have to be a fully functional backbone geographic/geometric and/or attribute database server system - as ray of discrete and disconnected "operational units" working autonomously (local intranet and/or local area network) - which are based on an "optimized" Total Cost of Ownership (TCO) at least at the woreda (district), regional, and federal government levels already in place. These have to be reliably standardized. The scalability, security, replication/(proper data synchronization), data consistency/integrity, data warehousing (which will support the decision process needs), and the high availability have to be guaranteed. The data entry points have to be clearly defined and maintained. Since the telecommunication infrastructure is not every where available and at some places not reliable, data transfer/actualization/replication among these servers and "operational units" may be securely done by implementing a policy of using such data carriers as CD-ROM from the clearly authorized source units "publishers" to the target units "subscribers" in a regular, timely, and routine manner.

The proper introduction of information technology, GIS included, may help in securing a reliable, optimal planning and transparent execution of the economical, social and environmental processes which may contribute decisively to the fulfillment of the above discussed government program. However, detailed policy actions with regard to information technology implementation measures needs to be clearly spelled out and the public at large have to be well informed, including an assessment of which applications and utilities have the greatest impact, in order to help prioritize them. Beyond that, the opening and operation of a dedicated computer/electronic technology college/institute would be the appropriate answer to the overall huge information infrastructure deficit of the country.

8.3 Measuring the Value of GIS and its Necessity for Ethiopia

As discussed in the preceding chapters, Geographic Information System (GIS) offers a unique opportunity to analyze and compare disparate types of information. Digital/analog cartography is inherently included in it. The ability to integrate traditional databases with geographic or spatially referenced information opens up new opportunities to deliver both information and services. The utility of GIS can be seen in applications, which range from protecting our natural resources to identifying trends affecting economic activities, to managing physical infrastructures. The important programmatic areas namely the economy, the environment; the community health, safety and security share many common information needs. While some information requirements are specific to one program area, there is an extraordinary amount of overlap among them. Much of the data needed for economic development are the same data needed for environmental conservation activities and for disaster preparedness and mitigation. In general terms:

- it allows more effective development and evaluation of policies to support economic growth and environmental protection,
- The technology will allow an increased ability to assess potential disasters such as drought, flood etc. - to mention some, which are frequent in Ethiopia - and create a mechanism to possibly avoid or at least alleviate their effects on the people and the livelihood of the rural community,
- it allows the integration of information from diverse sources in order to strategically develop a solid information infrastructure, and
- it can support new services that would not be feasible or even impossible under manual processes, due to its processing speed and better integrity.

For an effective utilization, GIS technology has to be internalized into an organization daily activity and be implemented by potential users. The three criteria often used to judge implementation success are user satisfaction, system usage and system performance. GIS technology cannot be said to be "properly implemented" if it does not lead to an increase in productivity, and environmental effectiveness for the organizations which apply it. Further reading is given in [Kohl1998].

The effective creation, maintenance, and sharing of spatial data across governmental and other agencies can serve a variety of vital private and public purposes. Such a collective approach may minimize the expense and allows more rational use of resources and professionals.

8.4 Barriers to Information Sharing and Coordination

In the implementation of a GIS, the main investment factors are the hard- and software expenses as well as the manpower in the creation and maintenance of the database. Effective application of GIS requires high skill in the different disciplines and in computer science. For Ethiopia, with its limited amount of fund availability, collective use of the GIS technology may bring higher benefit. GIS technology is in a dynamic development, with a short life-cycle often as short as few months and frequent upgrading demands. The computer hardware technology development shows a similar trend as well. These factors needs a nearer and continuous monitoring since they will adversely affect any meaningful implementation in the country.

There are some institutions in Ethiopia which have started to use the GIS technology. The individual introduction requires higher amount of starting expense at the beginning of the realization for the hardware, software, its selection process, purchasing, installation and secured operation. The steady management and actualization of the database is also one of the crucial and costly duty to be done. The expenditures for annual fees, and regular updating expenses in the case of new release (one could expect at least annually) are additional burdens, which hinder the updated and effective use of it.

The very similar nature of considerable amount of data and information for different institutions and state agencies, would cause unnecessary redundancy in effort and activity for producing the same type of result in different departments, institutions and agencies. This, basically should be avoided where ever possible. The benefits of sharing data are obvious. Sharing reduces the total cost of individual applications and can make GIS affordable for a wider spectrum of organizations.

Data sharing, however, is not as easy as one might think. Availability, pricing, and ownership are common bottlenecks. In some cases, data sharing is limited by the use for which the data were originally created. The scale or accuracy of data required by one organization may not be sufficient for use by another. Nevertheless, many data sets could be used by multiple organizations and a steady data standardization may allow the accuracy and quality management, [Kelley1995].

Some of the main management and policy factors, which hinder the sharing of spatial data in Ethiopia are:

i. lack of awareness of existing data sets,
ii. lack of or inadequate metadata,
iii. lack of uniform policies on access, cost recovery, revenue generation, and pricing,
iv. lack of uniform policies regarding data ownership, maintenance, and liability,
v. lack of incentives for sharing,
vi. absence of tools and guidelines for sharing, and
vii. absence of a federal-level guidance and leadership.

The focus on geography to create a comprehensive database acts as a principal motivator and is a distinguishing characteristic of multi-participant GIS programs. A shared geographical database brings entities together in multi-participant GIS program. Eventually, the convenience of common geography brings diverse participants together. Some other driving forces that lead to the creation of a multi-participant GIS are:

i. budget constraint,
ii. the inherent similarity of the information base,
iii. increased efficiency and cost-effectiveness,
iv. data interdependence,
v. new technology, system advancement and its integration,
vi. general awareness, demand and mandates among government organizations, and
vii. the intention of reducing costs and increasing efficiency.

In addition to the above factors, programs may be designed to facilitate sharing of responsibilities and funding. All these driving forces may motivate the interested institutions to come together to achieve their objectives.

In this regard experiences from different countries and organizations may give a vital input. Such an example could be the "Working Committee of the Surveying Authorities of the States of the Federal Republic of Germany, (AdV)". Its detailed activity could be taken from its homepage: http://www.adv-online.de. There, subject related additional links to several other governmental and nongovernmental institutions are available.

8.5 A Multi-Participant GIS Program and its Implementation Strategy

In addition to its technical and information handling capabilities, GIS is capable of enabling numerous organizational and economic implementations. This potential of GIS may lead to emergence of the multi-participant approach. In multi-participant program each participant has a unique culture, structure, policy, decision-making rule and expectations from implementation of GIS technology. A successful program depends upon aligning different characteristics of the program's structure to those participants, individually and collectively. Maximizing the system's potentials necessitates a well-defined implementation framework that can help manage changes and integrate the technology in organizations, [Kelley1995].

A multi-participant GIS program can be understood as a project that involves more than one user, each of whom have a different reason for implementing a GIS with responsibilities shared among all. The multi-participant system could be designed to perform applications and satisfy requirements of all participants. The involved participants may consist of a wide range of legally separated entities, all bound by common GIS needs. These participants may be:

i. one or more local or federal government departments and agencies,
ii. planning institutions and agencies,
iii. utility departments (i.e., electric, water, road, health, etc.),
iv. private sector agencies, Non Governmental Organizations (NGOs), and
v. many combinations of the above.

A multi-participant GIS program implementation will cross established organizational, departmental boundaries and can involve cooperative efforts amongst all participants. Also, these programs may involve participants, each with its own set of goals, functions, data requirements and reasons for implementing a GIS. They may differ in their management philosophy, culture and stage of technology use. Multi-participant GIS program may thus, involve diverse local entities which come together with unique motives and work together to achieve individual missions.

The concept behind a multi-participant approach to GIS is to share technology, costs, and the responsibilities without compromising the needed information quality of the participating agencies. It may help participants to obtain a fully operational system at a lesser cost than if they each had to develop the system individually. This will be even a more important benefit for the governmental agencies and those small organizations which would not be capable to develop it on their own in the first place. The development and maintenance of a comprehensive "alive" database - a diverse and expensive task - can be easily achieved through the multi-participant approach. The comprehensive database provides participants with data for carrying out functions and performing analyses that were not possible, as data was unavailable. The centralized database ensures consistency in accuracy, scale and format, and provides a central location to organize, store and maintain data and information. It eliminates redundancy of efforts and funds for collection, development and maintenance. The most important benefit from a multi-participant approach may be in terms of increased efficiency, productivity, and reduced costs for all participants,[NorthCarolinaDENR1999].

From the above discussion in this chapter, the following GIS implementation strategy can be forwarded:

i. advance and encourage the national spatial data infrastructure, adapt and promote a standard of digital geospatial metadata for data exchange and other purposes,

ii. enhance the nation-wide organizational framework, clarify roles and responsibilities for private, state and other agencies,

iii. develop and implement standards and procedures for creation and maintenance of a centrally coordinated GIS and attribute database, assign maintenance responsibilities,

iv. establish policies and procedures for access to and distribution of the geographic database, these have to address costs, legal issues, and procedures,

v. provide training, technical support, production services, ongoing promotion, education and evaluation,

vi. develop and implement a regular GIS utilization training for ministries and other state agencies, and

vii. develop policies for data development, data access, data security, and data quality.

As discussed above, a properly introduced GIS technology can build a solid information foundation which may be used for the countries overall balanced development in general and the government's short and medium term environment, water, economy, and social related plans and projects in particular.

9 Results and Discussion

Geological Composition and its Permeability

The present day landform of the area is mainly characterized by the epirogenetic uplifting of the early Tertiary, Eocene to early Oligocene. The massive trap basalt eruption in the Oligocene/ early Miocene, the rift valley formation due to the stress caused by the ongoing triple divergence drift of the Arabian-, Nubian-, and Somali- plate and the related volcanic activity - with its characteristic complex pattern of narrow belts of parallel horst and graben form - dominates the present day picture of the area. The renewed rift movement associated with the Aden volcanic series in Pleistocene, had shaped its most recent geomorphology, predominantly in the rift valley floor, further.

The volcanic eruptions and rift movements have supplied a huge amount of basalt, ignimbrite, scoria, tuff and other pyroclastic material to the surface. The trap basalt material exists predominantly as a series of sub horizontal lava flow, with variable size and intensity, spread over a large area. Ground water availability in such strata is primarily dependent on the weathering condition between its columnar joints. Ignimbrite is the most common rock in the region, both on the escarpments and the floor of the rift valley. It occurs as a successive stratum that vary in thickness. Some ignimbrites are separated by a paleosoil. At other outcrops, ignimbrites are interbedded with layers of graded lacustrine deposits made up of cemented sand, mud and clay. Intensive weathering of ignimbrites may transform it into a widespread impervious layer. Pumice is the other volcanic product found throughout the investigated region, and it generally occurs as a widespread bed without any welding or connection. The pumice comes often from a distinct volcanic center either as air-falls or pyroclastic flow.

The Quaternary deposit of the study area is characterized by a complex form of interbedding and/or interweaving of the volcanic and fluvial/alluvial sediments. The pluvial sediments in the rift floor widely consists of lacustrine and fluviatile clay, gravel and diatomite while in the interpluvials, loess and aeolian sand deposits were predominant.

The geological map of Nazereth, EIGS 1978, shows that the Pliocene/Pleistocene fresh alkali and per alkali rhyolites, trachytes, domes and flows constitute the most impermeable material of the area.

Climate and Rainfall

The monthly mean yearly average rainfall for the station Addis Ababa is 100 mm, followed by 76.3 mm for Debre Zeit and 71.4 mm for Nazereth. The yearly average mean maximum temperature of Nazereth, Debre Zeit and Addis Ababa are 27°C, 26.2°C and 22.6°C respectively. Where as their yearly average mean minimum temperature amounts to 14.2°C, 10.3°C and 9.8°C. This shows that the maximum temperature of Debre Zeit asymptotes that of Nazereth and its minimum temperature nears the Addis Ababa temperature. As a result, a higher temperature fluctuation is observed in Debre Zeit than in Addis Ababa or Nazereth, [Figure 9b].

The monthly mean minimum temperature peaks amounts 17.2°C for Nazereth in May/June, 11.6°C for Addis Ababa in May, and 11.9°C for Debre Zeit in June. In average, the Nazereth mean minimum temperature is approximately 4°C warmer than the other two. Even though there is higher fluctuation, the tendency is similar for both Nazereth and Debre Zeit. The monthly mean yearly average minimum temperature for Addis Ababa had increased from an average of 9°C in the 1950ies to an average of 11°C at the end of 1980ies, [Figure 10b]. The cumulative effects of different social processes, may have led to the increase of the mean minimum temperature. This may indicate the likely situation of microclimate change. Natural bio-diversity, both wildlife and vegetation may be at risk to diminish further, partly due to the lack of awareness and information about this evolutionary environmental change.

In applying the climate zone Kola (below 1800 meters a.s.l.), Woina Dega (between 1800 and 2500m. a.s.l.) and Dega (above 2500 m. a.s.l.) we can see that around 75% of the study area is in the Woina Dega and the rest in Kola climatic zone. Only few mountain tops are in the, nominal, Dega climatic zone, see [map19].

In the Kola climatic zone the pan evaporation is over 400% of the precipitation as shown in [Table 1] . Even in the Woina Dega areas such as Addis Ababa and Debre Zeit, the yearly average monthly mean value of the pan evaporation is 144 mm and 145,8 mm, respectively, whereas their respective value of rainfall amounts to 97.1 mm and 66.4 mm as shown in [Table 2], and [Table 3]. In the highland areas, (on the plateaus and escarpments) runoff is very high.

The monthly mean yearly average value of the sunshine for Addis Ababa and Debre Zeit is 6.7 hrs and 8.2 hrs respectively. The recently observed, steadily increasing investment activities in several industrialized countries show the economical viability of using the sun radiation as an energy source, for example see the link www.energie-online.de/links.html. Ethiopia with its high sunshine hours and intensity - which mainly results from its geographic location and the high altitude - obviously have a promising potential to use. The development of solar energy industry-sector in Ethiopia may ultimately decelerate the demand for firewood inducing a recovery of the environment. In addition to the traditional rain distribution analysis, institutions should give focus to this resource and its usage as an energy source. In the coming years, this abundantly available energy and light source should be taken into account by the responsible governmental agencies more seriously and industries of solar technology should be encouraged in different ways in order to increase their investment in the country.

Remote Sensing

The decision as to the most appropriate satellite data is mainly dependent on the study objectives, scale of the subject under study, temporal and financial constraints, and the nature of the anticipated results. In many instances, the use of multiple data types (multi date images, multispectral images and spatially enhanced images) will substantially aid the study. A number of these concepts reduces to the basic constructs of resolution and scale, each complexly and some what interdependently linked to the other. One couldn't address topics associated with either resolution or scale in isolation from the other, since they are interdependent quantities. The amount of data/information contained within an image depends in large part on the image resolution. Definitions of spatial, spectral and radiometric resolutions are not unique or single valued, since both technical and broader application-oriented usage exists for each. For a given remote sensing system, an improvement in resolution in the spatial, spectral or radiometric domains will result in some degradation in resolution in one or more of the other domains. In the selection process of image data, a total analysis should not be restricted to the engineering aspects of sensor design alone, but should be an integration of environmental engineering, economic, human, and social considerations.

In a data collection, often the most important point is the trade-off between the quality of data and cost of obtaining that data on the one hand, the scale and resolution of the expected result to-extract on the other. For any system selected, there is a residual level of error within the data beyond which it could not be economically justified to attempt for an improvement. On the other extreme side, it may not be necessary at all to have a high-resolution data in the first place for the specific problem-setting at hand and a "within scene noise" may surface causing artifacts.

For this study, digital Landsat TM, MSS and SPOT image were bought. The main criteria for selecting these data was the minimum cloud coverage. The form of the grey level distribution against the number of grey level values shows the resolution contrast of each band. [Figure 17] on page 47 and [Table 7] and [Table 8] on the pages 48 and 49 shows that the correlation of the neighboring bands of both Landsat TM and MSS is high, this is further elaborated in the appendices 1 and 2.

In the IHS-Domain saturation stretched and back to RGB domain transformed TM (4,7,5) in RGB gave a more contrast rich map, [map1]. In this map the dominant geomorphologic setups, surface water distribution, highly denuded escarpment regions and the rift valley itself are clearly depicted.

In [map2], the result of maximum likelihood classification of the principal components (3,1,2) from the Landsat MSS is presented. This map is additionally overlaid with the vectorized main river course as well as water shade vector of the full scene area derived from the 1:2 million scale original Ethiopian hydrogeological map.

On the [map3] and [map5], farming plots are detectable. These are partially fallow, harvested fields, meadow, and in some places farms of chick-pea. The escarpment and highland area is highly representative of this observation. This reality shows how extensive the land use in the escarpment and highland area is.

In the [map3], [map4] and [map5] the results of different principal component processing method are presented. The PC1 provided a measure of overall albedo, illumination and brightness. The substantial positive loading of PC1 is manifested in its ability to best reflect the geomorphologic structural setup of the study area including man made structures. The color composite [map4] gives the best color composite false color image of the area in relation to land use (mainly at the level of detection).

In both [map3] and [map4], the natural or related forest is clearly contrasted against the other background objects with a deep red color, with some variation due to the presence of PC1, the predominant morphology component.

On [map4], the contrast of the horst and graben formation, the series of circular volcanic eruption centers, and the few crater lakes are shown. Some of the volcanic centers show how far the outflow is moved from the center of the eruption. The retrogressive erosion extension, at the NE edge, can be clearly identified.

Textural filters have been applied on the Landsat MSS and TM scenes (original bands and neochannels) in order to create images containing enhanced geological information. Structures and textural signatures enhanced through gradient and edge detection filtering are described and put in relationship with the local geology. The resultant image from the 101x101 kernel-processing was less grainy than those results with 51x51 matrix-kernels. The convolution result with 51x51 convolution matrix, as presented in [map7], shows the recent sedimentation area structure more clearly than those with 101x101, [map6], which simply smoothed it.

The band ratioing method had delivered similar results like the convolution with respect to the regional setup and tectonical structure of the area. The [map9] and [map10] show the color additive color composite Landsat TM and MSS ratio image, the latter one, overlaid with the main identified lineament trends.

It is noticeable that certain areas seem to be less affected by the faulting and fracturing than the others, and that some appear completely unaffected. These facts are related to the nature and the age of the various terrain present in the region. More precisely, in large areas of the rift valley and on part of the escarpment, blankets of recent incoherent alluvial and lacustrine deposits obliterate the underlying bedrock picture while in others, the reduced densities of faulting and fracturing can be related to the fact that the rocks in these areas are of recent volcanic eruptions.

The discontinuities apparently offset parts of the rift floor, which, moving northward, appear displaced toward the east. These formations may have a key economical and environmental significance, they may be responsible for the underground water movement. In this regard, a further higher resolution study using geophysical methods such as the geoelectric and electromagnetic, may convey us clearer information about the discussed geological structure.

The relatively coinciding results of the geomorphology and structural fabrication of the study area had brought even more information than it was given by the geological and hydro geological mapping of the area. A further comparison was made with an original small digitized cutout of the geological and hydro geological map of the Nazereth area, with an original 250000 scale map, [map11] and [map12].

After this comparison of the newly processed lineament map with the above cutout, it is shown that:

- the newly created map has enhanced the information on the lineament distribution considerably,
- the main volcanic centers and chains of volcanic cones are mapped clearer on the new map,
- it was found that no lithological mapping could be made with the result from the unsupervised classification, and
- it was also seen that the overlay of the village distribution on this map has brought a much clearer understanding of the population distribution pattern than ever before, [map16].

For the study of the semi local to regional formations (structures), the Landsat MSS had delivered good result. This is, instead of the frequent prejudice against the Landsat MSS, showing the vital role it may play in delivering regional structural information.

It is found that texture analysis is efficient as far as tectonical lineament recognition and mapping is concerned. The choice of the orientation and size of the filters depends on the content of the original image (type of landscape, vegetation), the latitude (solar azimuth) and the date of recording (solar elevation). The processed spectral band or combination of bands were chosen according to the features to be enhanced. As a result of their relative independency towards the vegetation cover, their expression of the morphology and their capability to discriminate soils according to their moisture. The Landsat TM bands 4, 5 and 7 are the most suitable for the study of the structures and textural signatures in general. But since almost the whole area is covered with thick soil and actively used for agricultural use, it was difficult to map any lithological boundary in the area. Only the locations of intrusive magma and recent volcanic eruption cones can be distinctively associated on the resulted maps.

The NE-SW lineament structure is the predominant direction followed by the NW-SE. In the rose-diagram [Figure 18] on page 71, the dominant stress intensity lies around 30°, further discussion is given on pages 69-72. The exact lateral and vertical extension of this structure should be a topic of further research and investigation in the future. Especially, large-scale geophysical mapping methods such as resistivity, Very Low Frequency (VLF), electromagnetic, and partly refraction seismic methods are good combinations for mapping the structure of the shallow underground.

The results of the above discussion can be summarized as follows:

- increasingly more active axial bands of the rift, generally bounded by long linear features were outlined by tonal and textural variations, and
- discontinuities along such bands are marked by transversal linear features connected with the rejuvenated sectors of the transcurrent faults, delimited by long tectonic features and marked by volcanic center alignments.

Different enhancing methods involving linear combinations, ratios and logarithmic transformation of TM spectral bands aiming at differentiating and eventually quantifying the vegetation and soil, had been designed, adapted and subsequently used as an input for the classification. Training polygons had been selected according to a digitized general geological map and personal information at the time of the ground truth collection. Both supervised and unsupervised classification methods were performed and studied. The use of a classification method for identifying the lithologies and their subsequent boundaries was not promising as is shown in [map8]. The spectral information of the underground geology was substantially masked by the upper soil, vegetation and farm plots. The lithological identification and mapping on the other hand, did not bring any meaningful result contrary to the common expectation.

More refined land-cover classification schemes require more spectral and ancillary data such as antecedent conditions and crop calendars. It will be of great importance to follow this direction and map the land use of the area using higher resolution, for example 2-5 meters pixel size, multispectral sensor, which will certainly bring detailed quantitative information about the study area.

The combined result of SPOT and TM have resulted in a considerable geometric resolution gain. In this combination, some of the rural villages can be better indicated. The small pieces of farm lands can easily be detected on this merged image due to their local variation of dark red, dark blue light green colors, but they are too small to identify and map them. They tend to obliterate the further use due to their tendency of being a "within scene noise". Further, it showed a substantial village pattern change as a result of the villagization process in the mid 1980ies in the rural area. However, this detected dwelling pattern couldn't be identified and mapped.

The population distribution from the topographic map of the 1995 compared with the result processed from the SPOT shows inconsistency. Even though there was a substantial "local resettlements" because of the "villagization" activities of the 1980ies, the topographic map from 1995 did not take this fact into account. It contains a considerable amount of outdated village distribution information. The availability of an updated data is a prerequisite for better use of the resources and the tackling of the recurring drought and other social problems.

Due to the bad contrast of the SPOT data the expected identification and mapping of features from it was compromised. The result of the SPOT panchromatic image processing was below expectation which is mainly caused by the low contrast ratio and less background contrast of the study area. However, since the resolution of the SPOT panchromatic is considerably high, further temporal and spatial investigation shall be done prior to generalization of this conclusion. It is to expect that 5 meters or higher resolution satellite images from companies such as IKONOS or aerial photograph may enable in filling the above mentioned information gap and a further study in this direction is mandatory.

The major part of the aerial photograph from the Ethiopian mapping agency could not be considered for further detailed interpretation due to its poor quality in most cases.

GIS

In implementing a GIS for project-specific environmental analysis work or for more-programmatic uses, various issues have to be considered up front. These can be targeted by assessing user needs, the degree of geographic information culture - and the kind, scale, resolution, and volume of the data needed. A thorough and early assessment of the users' needs for data will help to ensure the development of an efficient operation. The supply and use of geological or any environmental information is a whole process that starts with the availability of raw data and proceeds to the computing machines which collate, treat, aggregate and analyze the data and produce the intended information for the users who rely on it for making their decisions. The number and type of users will influence how the data is collected and how the information system is to be designed and managed. An analysis should be made of the existing and proposed activities within the relevant agencies that rely on spatially referenced data, both graphic and non-graphic. Systems should be built on existing capabilities as much as possible. While it is impossible to define a single GIS blueprint for all environmental assessment and review, a number of basic characteristics for such a system is apparent.

The effective use of GIS can be seen as a vital information infrastructure tool for managing and handling the different inter related works and there by allowing a flexible analysis possibility without affecting the specific interest and condition of the participating institutions. Proper application of GIS may be a vital means for planning, implementation and post construction management. The flexibility, timeliness and versatile applicability of GIS is proven. In the development and application of GIS in Ethiopia, the internet technology should play a substantial role.

Generally there is an economy of scale which can be gained by developing this GIS database application system for further local use in the study area. It is a system consisting of a comprehensive database that can meet the needs of a variety of users. The database can be used not only for special projects, but also for any ongoing tasks in that specific geographic area. Therefore, this work can be seen as an extension of efforts to see the possibility of applying the technology to potentially cover the whole country and produce timely, scalable and reliable information. Proper and timely information is decisive to initiate or guide development projects – be it at the local, regional or federal level. Towards this end the GIS can deliver a substantial contribution.

By integrating the results of the image processing and the topographic maps, a GIS database with a total size of 5 GB was built. Using this database, it was possible to make versatile analysis by applying different overlays and techniques. Few of the possible results were presented in this work in form of maps. Even though not versatile enough, this database can be seen as an archive that must be expanded and maintained through time, and may be called upon to meet a variety of applications. The information that is generated through data examination and analysis is intended to result in a decision that may initiate or suspend an action. These can include archival databases for general reference (e.g., surface/ground water, and regional land use), as well as, analytically oriented databases or models that describe specific aspects of a large complex undertaking (e.g. erosion, protected drink water area).

The geology of the study area is mainly determined by the Mesozoic lava outflows and the subsequent rifting of the area. In the rift valley itself, there are irregular volcanic cones mainly from Quaternary volcanic activity. 44% of the study area is flat, from the rest 25.4% of the area have a slope up to 3° of inclination /declination. Further detail account is given in [Figure 23] and [Table 14]. The geomorphology of the area is a result of complicated and interleaved processes of tectonic and volcanic activities as well as exogen dentritic erosion and deposition.

By dividing the whole aspect into 8 aspect classes, the most frequent aspect in the study area is SE with 20.28% of the total count and this covers 20.15% of the total surface. Detailed account is given in [Table 15] and [Figure 24]. In every aspect interval, the slope is represented in all its inclination angle from 0.3° to 89.9°, revealing no predominant aspect correlation with certain range of slope angle.

The black and white representation of aspect [map23] enhances the contrast of the study area surface for better visual understanding. The shade index representation of [map20] delivers a good visibility of the geomorphology of the study area and gives a supporting information for the image interpretation which was discussed in the past chapters in detail.

In the escarpment area, in average there is a decrease of 1 meter elevation in every 20-40 meters moving towards the rift valley. Over 35% of the area with more than 30° slope angle are densely populated prompting a high erosion vulnerability of its use as a farm land.

The villages are almost evenly distributed throughout the study area. Intensive villagization is observed in the surroundings of the cities and towns, see [map16], [map19] and [map23]. In the north eastern and central parts, especially the village pattern tend to follow the horst formation of the geomorphologic pattern of the respective locality.

The diminished amount of forest, non availability of commercial wood and unproportionally high destruction of the existing “wild natural” vegetation should be addressed. In the short-term, a commercial forest may be meaningful for the supply of the villages and towns. In the long-term however, we have to go more and more away from firewood usage. In this regard, the local authorities must encourage the plantation of indigenous trees by the rural community and enhance environmental preservation in order to eventually enforce a steady mitigation of the natural-forest dwindling. Properly studied and well planned adaptive policy on solar, small-scale hydroelectric, and wind energy should be developed and implemented to encourage private companies.

The following factors can be stated as the three interrelated crucial environmental problems - the result of negative cumulative impact of the social processes - namely:

- The high demand of the villagers and local towns for fire-wood, as well as the increased pressure for more farm-land causing a rapid deforestation process and thereby creating a permanent damage on the environment,
- increased and routine use of high-inclination and environmentally sensitive locations as farm and grassland, the degradation of agricultural and grazing fields due to continuous and rapid run-off surface water in the rainy seasons, and
- destruction of the local micro climate, and subsequently, the elimination of the native natural flora and fauna.

The merit of introducing GIS and digital cartography in Ethiopia is indisputably high. In order to have effective use of this rapid developing technology, the collective approach depending on the geographical location and thematic similarity would be the best one.

Generally, the fast high-tech advance in hardware technology together with the development of professional GIS software, dataware and powerful relational databases allows us an ever better and real opportunity for a much detailed analysis, reconstruction and modeling of paleogeographic processes. This allows a better information management and coordination in all geo- and environmental science disciplines. The interdisciplinary information extraction such as the genetical, structural, lithofacies, and historical - from the geological studies can be easily integrated with the applied fields such as hydro- and engineering geology. The documentation of a given hydro geological site or a drinking water well can be easily and effectively accomplished. The digital information availability allows the end users better transparency and effective management of any environment related project.

The developments in the area of world wide web, internet, digital multi-medial data capturing, online services and publications are the other aspects of the merit of the digital world. The increased availability of electronic mass-storage devices with ever better performance and low price permits the further integration of multimedia information to the GIS database from which the country may benefit a lot.

Water

The identification of lineaments with their semi regional extension, has important ramification in applied geology, because such features can represent faults and fracture zones. These features in turn can signify potentially hazardous or economically important environments. They can serve as conduits for ground water at the present time. A lineament can also be identified as a seismically risky location, due to the possibility of subsidence caused by the rapid withdrawal of ground water or tectonical movements associated with seismic activity. The basin and drainage pattern along the study area is irregular and locally very variable which is a manifestation of the underlying structure of the area.

Faults may create a hydraulic connection between the aquifers, provided the displacement is sufficient. In the escarpment area, there are a number of faults that might allow water to flow from the upper into the deeper aquifer horizons and eventually to the rift valley aquifer horizonts.

The integration of GIS with the remote sensing has allowed the verification of some ambiguous locations. As examples could be taken the maps [map15], [map16], [map17], and [map23] for their distinctive rendering of the structural/morphological information. The ambiguities which may arise in the image processing - due to the indifference of the relief - for example in [map4], [map5] and [map9], can be clarified if one takes the supporting results from the maps [map20], [map23], and [map26].

The maps [map21], and [map24] give a good sculptural representation which allows a detail understanding of the area from different perspective angles.

Best result in detecting and identifying lineament structures delivers the maps, [map8], [map9], and [map10]. The [map6] also gives a substantial detail.

Presently, there are no major dam sites, water accumulating reservoirs or protected drink water areas, either on the plateau or on the escarpments. The absence of such water regulating mechanism is causing a constant insecurity of drink water and an accelerated depletion of the fertile soil due to erosion.

Unless there is a substantial effort to create a macro economically meaningful water regulating policy, and a mechanism of environmental relief, the overall development activity will stay negatively affected in the long term. For the time being at least, in the majority of the villages, an irrigation will remain a remote dream.

In the lowland area, (the main rift valley), the building of a dam will not necessarily be an optimal solution since:

i. evaporation is extremely high,

ii. the majority of the villages are located not only in the lowland but also equally in the highland areas.

The alkali and per alkali rhyolitic and trachitic domes, which are found widely spread in the study area, are areas of high runoff. Those domes do have steep sides and massive rock structures, which allow little or no infiltration.

The shallowness of a fresh volcanic material beneath the earth and recent hot plutonite intrusions may result in mineralization of the lower water bearing layer. Due to the activeness of the rift valley, geothermal and mineralized water upwards movement is common and this mineralization may extend even to the shallow aquifers and bed rocks of potential dam sites.

The drainage of the study area is partly circular, in the areas of big volcanic outflow centers, partly parallel, clearly following the northeast-southwest tectonic pattern of the western escarpment of the Ethiopian rift valley. In some areas, a dentritic drainage pattern is observed which inherently shows less possibility of groundwater building and high surface water run-off. In Debre Zeit area, there is no explicitly formed drainage. The rivers in the central part of the rift valley are short and seasonal. All in all, the drainage pattern of the study area is complex and reflects the frequent structural variation after every few kilometers of lateral extension, inducing high uncertainty of information extrapolation.

The potentially best recharge places in the study areas are the escarpment regions. The escarpment regions are mainly faulted basalt and ignimbrites, which have internal horst and graben structures. They have rugged topography covered by scarce and thin vegetation. These basaltic and ignimbrite rocks are highly faulted and fractured so that they generally cannot hold water for long. Instead what expected is the water which percolates through these fractured rocks mainly flows underground to recharge the aquifers beneath the lacustrine sediments in the rift floor area. The escarpment region rocks, even though are the best recharge areas, are not good aquifers. A further hydro geological investigation – such as tracing – may give clearer picture of this interrelation.

The overlay of dwelling pattern on slope and aspect [map23] shows that often the people in the study area live on the topes and foots of the volcanic cones, on the top of the horst and on the plateau. This makes water management and supply mechanism difficult. Here, some dwelling adjustment may be necessary and the water availability must be taken as one of the main factors in the future.

Groundwater exploitation and water management as a whole should be considered as part of the ecological system management. The inseparability of ground water and surface water as well as land use should be clearer to the concerned parties and to the society as a whole. Towards this end further large-scale study and selective pilot project execution is recommended. The increased indiscriminate use of fertilizer, pesticide and other chemicals by the farmers are the other potential sources of environmental pollution. In this respect, at least in the areas of tectonical weak-zones and higher underground water circulation horizonts, an adaptive regulation have to be developed and implemented.

Drinking water availability and the whole economic activity of the study area is totally dependent on the hope of a timely, regular, evenly distributed, and enough seasonal rainfall. Any future balanced resource utilization and environmental management should address the local energy consumption demand, and the necessity of water recharge reservoirs along and within the escarpments. A well preserved environment is the only security for the future existence of durable and healthy rural agriculture. This fundamental interrelation – between society and environment - should be properly understood by the responsible persons and awareness should be substantially increased among them.

10 Conclusions and Recommendations

Conclusion

The meteorological data from the Ethiopian meteorological agency was interpreted and the result was integrated with the built GIS database to analyze the rain, sunshine, pan evaporation, and temperature distribution in relation to the geomorphology of the respective study area locations. The full scenes of Landsat and SPOT satellite images were first digitally processed. This processing includes contrast optimization, ratio, principal component and inverse principal component analysis, IHS and inverse IHS processing, convolution and FFT operations. Besides, six topographic sheets of 1:50000 original scale were scanned and integrated. Contour, village, river, forest, well and spring coverages were vectorized. A variety of maps, which depict the tectonical, morphological and surface water, dwelling pattern and distribution in the area were created. Maps of different scales and a rose-diagram using the spectral, spatial and geometric resolution of the satellite imageries were produced. Slope, aspect, drainage, dwelling pattern, forest, spring and well overlay were created. The slope and aspect overlays helped to understand, and better describe the topographic exposure of the terrain. Attempts were made to extract a maximum information with special attention to hydro- and engineering geological applications using different scale and interpretation techniques. Possible land use management and effective use of GIS technology in Ethiopia were discussed.

The interpretation of the meteorological data shows a relatively uniform amount of rainfall for the past 30 years, though there was intensity and rainfall time variation over the years, [Figure 8b]. The yearly average monthly mean minimum temperature for Addis Ababa showed an increase of 2°C in this time interval.

The complex stratigraphic situation of the area - due to the occurrence of volcanic layers and intrusions of different ages belonging to various volcanic centers, as well as pyroclastic, eluvial and alluvial sediments - made it extremely difficult to depict and extrapolate the hydro geological pattern of even quite small areas.

The hydrology/hydrology condition in the study area can be divided into two main categories. Namely, the escarpment area with its higher rainfall and subsequent high runoff - which often finds itself in water scarcity soon in few months after the rainy season - and the lowland areas with less precipitation amount, high pan evaporation and more likely possibility of under ground water recharge, at least, in parts of the rift valley area.

As to the regional lineament identification, mapping, and the regional geomorphological study, the Landsat MSS image gave good result. A reasonably high regional water circulation is expected at least along the fault zone, the openings of the columnar basalts and along their fracture and joint surface. The extent and intensity of these lineaments and their potential economic benefit shall be further investigated using hydro geological tracing methods as well as geophysical methods such as electrical, electromagnetic and shallow refraction seismic methods. A further, more detailed, study in selected feasible areas shall be done using well-logging, pumping and other hydro geological methods.

The main economic activity of the study area is private subsistence agriculture with a complete dependency on the seasonal and timely rainfall. The distribution of the village is irrespective of the climate zone. Considerable amount of the villages are located on the tops and sides of the horst formation and on the sides of the conic-shaped passive volcanic eruption centers. This obscures a feasible modern water supply and management in any future planning. Effective clean water policy may force to change a substantial part of the now existing village pattern, which took no such conditions into account before.

The absence of any meaningful and systematic small scale dam for the rural population left them to be fully dependent on a timely seasonal rain. Any future assessment process on water/environment/soil-conservation management in the area requires detailed information on the economic status of the inhabitants, the types of crops grown, and the responsiveness to incentives for these programs from the villagers. Further, selecting the appropriate projects and conservation/land-rehabilitation models requires information on the land capability and its suitability for different uses. In a large-scale study, the high resolution image data from companies such as IKONOS may be appropriate for detailed land use mapping and planning purposes.

The future development of the study area largely depends on how well the water in the rainy season is managed. Water security of the region is totally dependant on how successful we can manage to preserve the precipitation which we get within two to four months of time for a longer period of months, sometimes years in order to eventually overcome the absence of expected seasonal rainfall. This can be approached through new basin wide water management, which should include the escarpment and the rift valley area as a single ecological-unit and take into account the environment architecture of the area as a single entity. In this relation building “many separate” small dams – coupled with local hydroelectric generation - along the escarpment may be more meaningful and could induce a sustainable development for the study area.

Recommendation

The GIS data base in this work showed a very high data integrity, scalability and information management facility, which can lead each task to a well coordinated target oriented process. Such an information infrastructure can help in clarifying the negative impact of cumulative social process on the environment and its possible mitigation. For any future negative cumulative impact mitigation and meaningful environment management, digital information infrastructure is indispensable. Towards this end, GIS technology may play a critical role. The build up of key digital environment infrastructure should be considered as a compulsory, which needs the proper attention from the local and federal government agencies.

The recent short/medium term development program of the government shows an inherently heavy dependency on a seasonal, and timely rain. Unless an economically feasible and sustainable introduction of water/environment conservation/management and local hydro/wind/solar electric generation adaptive-policy for virtually each group of village is in place – based on the specific environment architecture (the geology/geo-morphology, native natural flora and fauna, climate regime and meteorology) of a given local area - the rural agriculture development would remain permanently dwarfed and the subsistence farmers may most likely be captured in a cyclic destitution.

Up to now, there is no detailed environment management policy in place, either from the local or from the central governmental institutions for the study area. Neither there is a policy for drinking water management, nor there is any protected water catchment area or surface water recharge location. In any new development planning, there have to be a regional agricultural utility and environmental preservation plan, which outlines and prohibits converting watershed as well as environmentally sensitive areas into farmland. A new culture of building water protection area, awareness of environmental balance and securing enough clean water in every rural community should be appreciated, for example by exempting from payment of tax and other attractive incentives.

The financial/technical capacity of the villagers in the study area - for any meaningful water/environment conservation undertaking - is very weak, while they could make a huge labor power on a short notice available. The demand for water and energy by them is increasing steadily. There is a clear disparity between demand and supply in this regard.

Based on the results of this study, an integrated approach to the surface/groundwater as a single entity is recommended. The approach of building "many small" dams more frequently upstream at the valleys - along the escarpments and plateaus is more preferable. This causes the deceleration of the erosion, and less evaporation. Additionally, the availability of water for the villages around these high land localities will be secured, the villages in the lowland area and in the rift valley can get clean water in form of ground water in their vicinity. In such a method, the risk of mineralization on the rift floor could also be confined. Besides this, a wide possibility will be opened for the construction and operation of several "decentralized and small" hydroelectric power-stations for a local consumption.

Concerned institutions should develop and implement a long term adaptive-policy first to mitigate the scarcity of drinking water and local energy base. Parallel to the mitigation some locations - depending on the environment and social considerations - from the plateau, escarpment and lowland areas could be selected and pilot projects on environment regeneration, rain water harvesting, groundwater and spring reinforcement, small size water reservoir construction and energy generation should be done. Depending on the results and experiences, multiple of such adaptive-projects and economically meaningful operations could be conducted. Since the study area is highly populated, the owners of the locations - which may be selected for such conservation - should get a sustainable and economically meaningful compensation. In a most practical case, those people could be coupled to the respective specific projects by integrating them as the partial owner and supplier of the locally generated electric energy as well as water. This can be seen and further encouraged as a sort of specialization and labor division among the villagers.

Literature Cited

[Abera1989] Teshome Abera - A Report on a process water supply option to ELMICO's Awash Marble & Granite Factory Ethiopian Institute of Geological Surveys Addis Ababa EIGS 1989

[Berhe1991] Debalkew Berhe - Einsatz von GIS und Fernerkundungsdaten zur Entwicklung einer kostengünstigen, ökologisch verträglichen Landnutzungsplanung am Beispiel Äthiopien. München, Univ., Forstwiss. Fak., Diss., 1991

[Billingsley1975] F. C. Billingsley - Noise Considerations in Digital image processing Hardware, topics in applied physics Volume 6 Berlin-New York Springer Verlag 1975

[Billingsley1975] F. C. Billingsley, Edited by T. S. Huang - Noise considerations in Digital Image Processing Hardware Topics in Applied Physics Vol. 6 Picture Processing and Digital Filtering Berlin New York Springer Verlag 1975

[Bohumir1982] S. Bohumir, K. Araya, B. Bekele - Geophysical survey for Hydrogeological study in the Nazereth area (Sheet NC 37-15) Ethiopian Institute of Geological Surveys Addis Ababa EIGS 1982

[Caldini1987] F. G. Caldini, M. A. Pisanello, V. Arno, F. Salvini, M. Ghigliotti - Structrual Geological and Geothermal quantitative interpretations of Satellite images in the southern Afar region - Ethiopia Ethiopian Institute of Geological Surveys – Geotermica Italiana sri Pisa EIGS 1987

[Chang1991] Chang, K., and B. Tsai. 1991. - The effect of DEM resolution on slope and aspect mapping. Cartography and GIS 18(1):69-77.

[Chuchip1997] K. Chuchip - Satellite data analysis and surface modeling for landuse and land cover classification in Thailand F. Voss Berliner geographische Studien band 46 TU-Berlin, Germany

[Colvocoresses1975] A. P. Colvocoresses - Platforms for remote sensors: in the Manual of Remote sensing American Society of Photogrammetry Volume 1 Falls Church, Virginia 1975 p. 538 - 588

[Colwell1983] R. N. Colwell, D. S. Simonett F.T. Ulaby - Manual of Remote Sensing second edition American Society of Photogrammetry Virginia USA The Sheridan Press USA 1983, P. 548

[Colwell1983_1] R. N. Colwell, D. S. Simonett F.T. Ulaby - Manual of Remote Sensing second edition American Society of Photogrammetry Virginia USA The Sheridan Press USA 1983, P. 721

[Eshete1982] Gebretsadik Eshete - Watter Supply Augmentation study of Debrezeit Flour Mills and Maccaroni Factory Ethiopian Institute of Geological Surveys Addis Ababa EIGS 1982

[ESRI1994_1] ESRI Inc. GIS by ESRI ARC-INFO Version 7, - The ArcDoc series, Network Analysis Reference Manual ESRI Inc. USA ESRI Inc. 1994

[ESRI1994_2] ESRI Inc. GIS by ESRI ARC-INFO Version 7, - The ArcDoc series, Map Display, Query and Output ESRI Inc. USA ESRI Inc. 1994

[ESRI1994_3] ESRI Inc. GIS by ESRI ARC-INFO Version 7, - The ArcDoc series, Cell-based Modelling with GRID ESRI Inc. USA ESRI Inc. 1994

[ESRI1994_4] ESRI Inc. GIS by ESRI ARC-INFO Version 7, - The ArcDoc series, ARC-INFO data management concepts data models database desingning and storage ESRI Inc. USA ESRI Inc. 1994

[ESRI1994_5] ESRI Inc. GIS by ESRI ARC-INFO Version 7, - The ArcDoc series, ArcStorm and Map Libraries ESRI Inc. USA ESRI Inc. 1994

[ESRI1994_6] ESRI Inc. GIS by ESRI ARC-INFO Version 7, - The ArcDoc series, ArcPlot Commands ESRI Inc. USA ESRI Inc. 1994

[ESRI1994_7] ESRI Inc. GIS by ESRI ARC-INFO Version 7, - The ArcDoc series, Editing coverage and tables in ARCEDIT ESRI Inc. USA ESRI Inc. 1994

[ESRI1994_8] ESRI Inc. GIS by ESRI ARC-INFO Version 7, - The ArcDoc series, GRID Commands ESRI Inc. USA ESRI Inc. 1994

[ESRI1994_9] ESRI Inc. GIS by ESRI ARC-INFO Version 7, - The ArcDoc series, Map Projections Georeferencing spatial data ESRI Inc. USA ESRI Inc. 1994

[ESRI1994_10] ESRI Inc. GIS by ESRI ARC-INFO Version 7, - The ArcDoc series, Surface Modelling with TIN ESRI Inc. USA ESRI Inc. 1994

[EthiopianMappingAuthority1990] - Ethiopian Mapping Authority Map Catalogue Second Edition Ethiopian Mapping Authority Addis Ababa 1990

[Evans1972] Evans, I. S. - General geomorphometry, derivatives of altitude, and descriptive statistics. p. 17-90. in R. J. Chorley ed. spatial analysis in geomorphology. Methuen and Co. Ltd., London, 1972.

[Fisseha1997] Shimelis Fisseha - Resistivity Sounding Survey for Ground Water in Gambela Region Ethiopian Institute of Geological Surveys Addis Ababa EIGS 1997

[Floyd1987] F. Folyd, Jr. Sabins - Remote Sensing Principles and Interpretation W. H. Freeman and Company NY Second Edition New York Remote Sensing Enterprises 1987 P. 134

[Goeth1981] A. F. H. Goeth, L. C. Rowan - Geologic remote sensing Science Volume 211 1981 4484 P 781 – 791

[Gillespie1976] A. R. Gillespie - Directional fabrics introduced by digital filtering of images Proceedings of the second international conference on basement tectonics Newark Delaware 1976 P. 500 - 507

[Gillespie1984] A. R. Gillespie, A. B. Kahle, and F. D. Palluconi - Mapping alluvial fans in Death Valley, California, using multichannel thermal infrared images Geophysical Research letters Volume 11 P. 1153-1156

[Hadwin1975] Peter Hadwin - Boreholes in Ethiopia Ethiopian Institute of Geological Surveys Addis Ababa EIGS 1975

[Hadwen1974] Peter Hadwen - Earthquakes-Hydrogeological Effects in Ethiopia Ethiopian Institute of Geological Surveys Addis Ababa EIGS 1974

[Hadwen1975] Peter Hadwen - Generalisation and Hydrogeological maps of Ethiopia Ethiopian Institute of Geological Surveys Addis Ababa EIGS 1975

[Hassouna1997] K. M. Hassouna - Developing a natural resource database for Geographic Information System (Master of Forestry thesis) Virginia Polytechnic institute and State Univ. Blacksburg, Virginia 1997 76 p.

[Herdman1994] R. G. Herdman, Rodney Nichols - Remotely sensed data: Technology management and markets US Government Printing office Washington, DC 1994 194 p. ISBN O-1 6-045180-9

[Hildebrandt1997] Marion Hildebrandt, - Waldmonitoring mit Hilfe von Landsat-TM- und SPOT-XS-Daten für einen Wald-Management-GIS in den Bale Mountains / Äthiopien, Unpublicated Diplomarbeit des Studiengang Kartographie an der TFH Berlin, 1997

[Imkemeyer2000] Olaf Imkemeyer, - Erarbeitung einer Statellitenbildkarte der Boorana Lowlands in Südäthiopien unter Verwendung von Landsat ETM+ Daten sowie vorliegenden topographischen Daten im ARC/INFO Format, Unpublicated Diplomarbeit im Studiengang Kartographie an der Hochschule für Technik und Wirtschaft Dresden (FH) Fachbereich Vermessungswesen und Kartographie, Dresden 2000

[Isaacson1990] Isaacson, D. L., and W. J. Ripple. - Comparison of 7.5-minute and 1-degree elevation models. Photogrammetric Eng. and Remote Sensing 11:1523-1527, 1990.

[Kahle1984] A. B. Kahle - Measuring spectra of arid lands, in Deserts and arid lands The Hague Netherlands Martinus Nijhoff Publishers 1984 P. 195 - 217

[Kahle1981] A. B. Kahle; J. P. Schieldge, M. J. Abrams, R. E. Alley and C. J. LeVine - Geologic applications of thermal inertia imaging using HCMM data Jet Propulsion Laboratory Publication Pasadena Calif. 1981 P. 81.155

[Karavezyris2000] V. Karavezyris - Prognose von Siedlungsabfällen. Untersuchungen zu determinierenden Faktoren und methodischen Ansätzen - Disputation: TU-Berlin, Germany 2000

[Kebede1982] Getahun Kebede - Borewell Siting for Misrak Flour and Oil Mills, Addis Ababa Ethiopian Institute of Geological Surveys Addis Ababa EIGS 1982

[Kelley1995] K. Kelly, T. Pardo, S. Dawes, A. DiCaterino, W. Herald - Sharing the costs, Sharing the Benefits, The NYS GIS Cooperative Project New York State Department of Environmental Conservation Center for Technology in Government Project Report 95-4 1995 59 p.

[Kenea1997] Nasir H. Kenea, - Digital Enhancement of Landsat Data, Spectral Analysis and GIS data Integration for geological studies of the Derudeb Area, Southern Read Sea Hills, NE Sudan. Berliner Geowiss. Abh., D, 14, Berlin 1997

[Koch1996] Wolfgang Koch - Analyse und Visualisierung Geowissenschaftlicher Daten mit Hilfe digitaler Bildverarbeitung und eines Geo-Informationssystems. Beitrag zur regionalen Geologie der Red Sea Hills, Sudan. Berliner Geowiss. Abh., D, 12, Berlin 1996

[Kohl1998] N. Kohl, - Project officer Environmental data management initiative, One stop grant project plan Minnesota Pollution control Agency Chicago, IL 60604 1998

[Lex2000] Andra, - Vegetationsklassifizierung von Satellitenbildkarten und aufbau einer topographischen Basis für ein Rangeland-GIS- Vegetationsklassifisierung Boorana Lowland/Äthiopien, Unpublicated Diplomarbeit im Fachbereich Bauingenieur- und Geoinformationswesen der TFH Berlin, 2000

[Lintz1976] J. Jr. Lintz, D. S. Simonett - Remote sensing of environment Adison-Welsi Publishing Co. 1976 694 P.

[List1992_1] F.K. List, P. Bankwitz - Basic principles of remote sensing. Proceed of the 3rd United Nations Int. Training Course on Remote Sens. Appl. to Geol. Sci., Potsdam and Berlin. – Berliner Geowiss. Abha., D, 5, 27-35, Berlin 1992

[List1992_2] F.K. List, W. Koch, and M. H. Salahchourian, - Geological mapping in arid regions of Africa using satellite data-integration of visual and digital techniques. –Int. Arch. Photogramm. Remote Sens., 29 (B4), 325-332, Washington DC 1992

[List1993_3] F.K. List, P. Bankwitz - Fundamentals of digital image processing for geologic applications. Proceed of the 4th United Nations Int. Training Course on Remote Sens. Appl. to Geol. Sci., Potsdam and Berlin. – Berliner Geowiss. Abha., D, 5, 7-29, Berlin 1992

[Mbiliny2000] B. P. Mbilinyi - Assessment of Land Degradation and its Consequences: Use of Remote Sensing and Geographical Information System Techniques. A Case Study in the Ismani Division, Iringa Region, Tanzania - Disputation: Berlin, Germany 2000

[McCarthy1956] H. H. McCarthy, H. C. Hook and D. S. Know - The measurement of Association in Industrial Geography; Report 1 University of Iowa Iowa City University of Iowa 1956

[McCombsII1997] J. W. McCombs - II Geographic Information System Topographic Factor Maps for Wildlife Management (Master of science in fisheries and wildlife Sciences thesis) Virginia Polytechnic institute and State University Blacksburg, Virginia 1997 141 p.

[Melaku1982] Berhane Melaku - Hydrogeology of the upper Awash Basin Upstream of Koka Dam Ethiopian Institute of Geological Surveys Addis Ababa EIGS 1982

[Meissner2002] B. Meissner (Editor) - Managing Natural Resources - Strengthening Regional Development in Ethiopia, publication in progress Berlin TFH Berlin

[Missengue2000] F. Missengue - Die Auswirkungen der Konzessionspolitik in der Holznutzung der Republik Kongo auf Ökonomie, Ökologie und Raumplanung, bezogen auf die Hauptstadt Brazzaville - Disputation: TU-Berlin, Germany 2000

[Mohr1971] Paul A. Mohr - The geology of Ethiopia Haile Sellassie I University Press Addis Ababa Central Printing Press 1971

[Moik1980] J. G. Moik - Digital processing of remotely sensed images NASA Scientific and technical information branch NASA 1980 330 P.

[Moore1978] G. K. Moore - Satellite surveillance of physical water quality characteristics In: International

Symposium on remote Sensing of Environment, Michigan Michigan 12th Proc., Ann Arbor, Michigan: Environmental Research Institute of Michigan 1978 p. 445-462

[Niestè] A. Niestè - Drought risk modelling in the Nile valley B. Hofmeister, F. Voss Berliner geographische Studien Band 39 TU-Berlin, Germany

[NorthCarolinaDENR1999] North Carolina DENR - Business plan Geographic Information System, Needs analysis American Cadastre, Inc. Volume 1 North Carolina Department of Environment and Natural Resources 1999 99 p.

[Ongsomwang] S. Ongsomwang - Forest inventory, Remote Sensing and GIS for forest Management in Thailand B. Hofmeister, F. Voss Berliner geographische Studien Band 38 TU-Berlin, Germany

[Orthaber1999] Harald J. Orthaber. - Bilddatenorientierte atmosphärische Korrektur und Auswertung von Sattelitenbildern zur Kartierung Vegetationsdominierter Gebiete, Kartographische Bausteine Band 16 TU-Dresden 1999

[Paola1970] G.M. Di. Paola - Geological-Geothermal Report on the central part of the Ethiopian rift valley Ethiopian Institute of Geological Surveys Addis Ababa EIGS 1970

[Pregitzer1984] Pregitzer, K. S. and C. W. Ramm. - Classification of forest ecosystems in Michigan. p. 114-131. Proceedings of the symposium. Forest land classification: Experience, problems, and perspectives. March 18-20. Madison, 1984, WI. 276pp.

[Pilger1976] A. Pilger - Afar between Continental and Oceanic Rifting, Inter-Union Commission on Geodynamics Scientific Report No. 16, Proceedings of an International Symposium on the Afar Region and Related Rift Problems held in Bad Bergzabern, FR. Germany, April, 1974 E. Schwizerbart'sche Verlagsbuchhandlung (Nägel u. Obermiller) Stuttgart, Germany E. Schwizerbart'sche Verlagsbuchhandlung (Nägel u. Obermiller) 1976

[Ramsch1997] J. Ramsch - Implementation des ATKIS-Datenmodells in GIS mit relationaler Attributeverwaltung unter dem Aspekt der Fortführung (Diplomarbeit) Universität Leipzig, Institute für Informatik Universität Leipzig 1997 138 p.

[Rao1981] A.S.M. Rao - Water supply augmentation study of Meta Abo Brewery and Distribution Centre, Sebeta Ethiopian Institute of Geological Surveys Addis Ababa EIGS 1981

[Rao1997] R. Rao - An approach to open space planning based on the principles of landscape ecology: An application to greater Roanoke area (Master of Landscape Architecture thesis) Virginia Polytechnic institute and State University Blacksburg, Virginia 1997 91 p.

[Rein2000] H. Rein - Ansatzpunkte zur Effektivierung der Landschaftsplanung am Beispiel der Landschaftsrahmenplanung Brandenburg - Disputation: TU-Berlin, Germany 2000

[Rösch1998] N. Rösch - Topologische Beziehungen in Geo-Informationssystemen (Dissertation] Universität Fridericiana zu Karlsruhe (TH) Karlsruhe 1998 94 p.

[Rösler1976] A. Rösler - Afar between Continental and Oceanic Rifting, Inter-Union Commission on Geodynamics Scientific Report No. 16, Proceedings of an International Symposium on the Afar Region and Related Rift Problems held in Bad Bergzabern, FR. Germany, April, 1974 E. Schwizerbart'sche Verlagsbuchhandlung (Nägel u. Obermiller) Stuttgart, Germany E. Schwizerbart'sche Verlagsbuchhandlung (Nägel u. Obermiller) 1976

[Sabins1983] F.F. Sabins - Remote sensing laboratory manual Remote Sensing Enterprises La Habra California 1983

[Schandelmeier1990] H. Schandelmeier, D. Pudlo - The central African fault zone in Sudan – A possible continental transform fault. Berliner Geowiss. Abh., A, 120.1, 31-44, FU-Berlin, 1990

[Schneider2001] Sabine Schneider, - GIS im Ressourcenschutz am Beispiel Integrierter Waldbewirtschaftung in Äthiopien, Unpublicated Diplomarbeit Humboldt-Universität zu Berlin Fachbereich Geographie, 2001

[Sidle1985] Sidle, R. C. - Factors influencing stability of slopes. p. 17-25. in D. Swanston ed. Proc. of a workshop on slope stability: Problems and solutions in forest management. Gen. Tech. Rep. PNW-180

USDA For. Serv. Pac. Northwest For. 1985, Range Exp. Stn. Portland, OR.

[Simonett1976] D. S. Simonett - Remote sensing of cultivated and natural vegetation: cropland and forestland Remote sensing of environment Addison-Wesley Publishing Co. 1976 P. 454 - 459

[Slater1983] P. N. Slater, F.J. Doyle, N.L. Fritz, R. Welch - Photographic Systems for remote sensing in R. N. Colwell manual of remote sensing American Society for remote sensing and remote sensing Volume 1 Second Edition Falls Church, Va. The Sheridan Press 1983 P. 231 – 291

[Slater1985] P. N. Slater - Suverey of Multispectral imaging Systems for earth observations Remote sensing of Environment Volume 17 1985 p.85-102

[Soha1978] J. M. Soha, and A. A. Schwartz - Multispectral Histogram normalization contrast enhancement In 5th proceedings of the 5th Canadian symposium on remote sensing Victoria, British Colombia 1978 p. 86 - 93

[Solomon1974] Igzaw Solomon - Report on Hydrogeological Study on Debre Zeit Area Ethiopian Institute of Geological Surveys Addis Ababa EIGS 1974

[Stone1972] K. Stone - A Geographer's Strength: The multiple-scale approach The Journal of Geography Volume 61 1972 No. 6 p. 354-362

[Tadesse1993] Tarekegn Tadesse, Kumela Deressa, Tadesse Dessie - A Report on Recent Land Slide and resulting Damages in the Blue Nile River Gorge and its Tributaries, Eastern Gojam Zone Ethiopian Institute of Geological Surveys Addis Ababa EIGS 1993

[Tadesse1981] Ketema Tadesse - Bore Well Siting for the Ethiopian Building Construction Authority Head Office, Addis Ababa Ethiopian Institute of Geological Surveys Addis Ababa 1981

[Townshend1980] J. R. G. Townshend - The spatial resolving power of earth resources satellites; Memo 82020 Goddard Space Flight Center 1980

[Tsehayu1990] K. Tsehayu, T. Hailemariam - Engineering geological Mapping of Addis Ababa Ethiopian Institute of Geological Surveys Addis Ababa EIGS 1990

[Vernier1985] A. Vernier, T. Chernet, H. Girmay - Hydrogeology of the Addis Ababa area Ethiopian Institute of Geological Surveys Addis Ababa EIGS 1985

[Wishmeier1965] Wishmeier, W. H., and D. D. Smith. - Predicting rainfall-erosion losses from cropland east of the Rocky Mountains. ARS-USDA in cooperation with Purdue University. 1965 Purdue Agric. Exp. Sta. Handbook No. 282. 15pp.

[Zewde1994] T. Zewde, M. Wubishet, S. Mekuria, T. Mengesha, G. Birusa - Integrated Geophysical Investigations for Dire Water supply dam project Ethiopian Institute of Geological Surveys Addis Ababa EIGS 1994

Appendix 1. Pair wise scatterogram of Landsat MSS scene 168/054 (row/path) taken on the 21-st of April 1984 for the central part of Ethiopia, which is composed of (3168x3161 pixels in the horizontal and vertical axis).

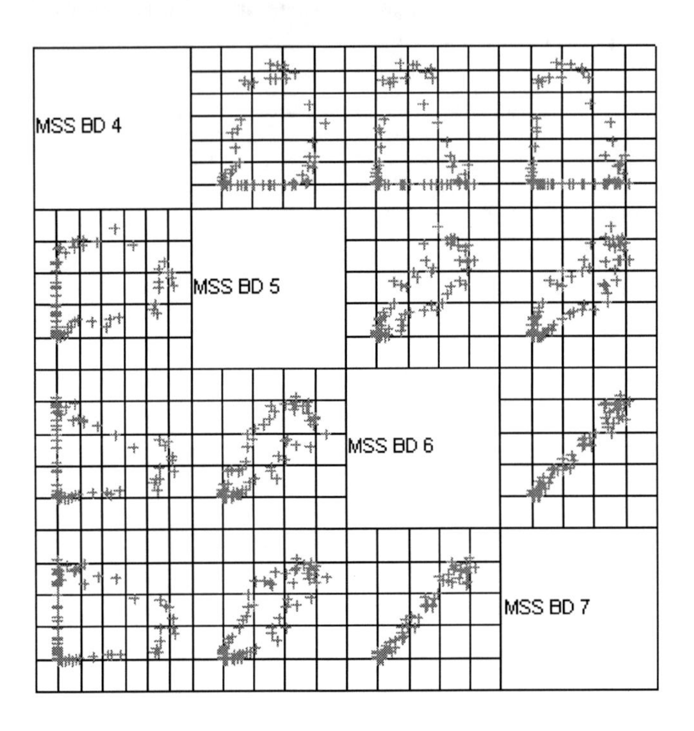
MSS BD 4
MSS BD 5
MSS BD 6
MSS BD 7

Appendix 2. Pair wise scatterogram of Landsat TM scene 168/054 (row/path) taken on the 5th of January 1986 for the central part of Ethiopia, which is composed of (7421x5964 pixels in the horizontal and vertical axis).

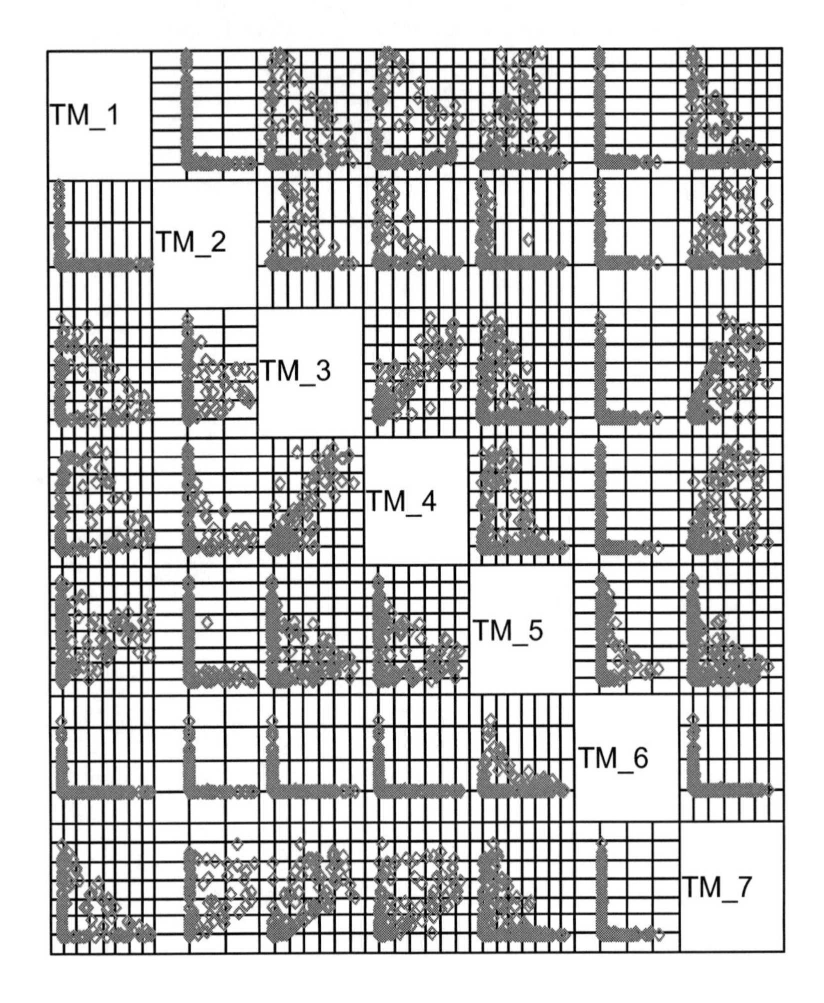
TM_1
TM_2
TM_3
TM_4
TM_5
TM_6
TM_7

Appendix 3. A program module written in the C-Programming language for transforming the ID-Values in the ARCEDIT coverage module into Z-values in the TIN moduledata-format of the Arc/Info.

```
/* Converts the ID-values of ARCEDIT into z-values of the TIN module */

#include<stdio.h>
#include<ctype.h>
#include<string.h>

main(){
 FILE *fpin,*fpout,*fpstat;
 char c,line1[50],line2[50],infile[20],outfile[20],statfile[20];
 short i;
 int j,count;
 printf("Give the input file name\t:");
 scanf("%s",infile);
 if(!(fpin=fopen(infile,"r")))
   {
    system("clear");
    printf("CANNOT OPEN FILENAME \"%s\" TO READ",infile);
    printf("\nCheck from the list below......\n\n");
    system("ls -la");
    printf("\nPROGRAM TERMINATED.. Re Execute\n\n\n\n");
    return(0);
   }
 printf("Give the output file name\t:");
 scanf("%s",outfile);
 fpout=fopen(outfile,"w");
 printf("Give the output statistics file name\t:");
 scanf("%s",statfile);
 fpstat=fopen(statfile,"w");
 for(i=0;i<40;i++){
   line1[i]=' ';
   line2[i]=' ';
```

```
}
j=1;
while(!feof(fpin)){
 switch(j){
 case 1:
   for(i=0;i<39 && (c=getc(fpin))!=EOF && c!= '\n';i++)
     line1[i]=c;
   line1[i]='\0';
   fprintf(fpout,"%s\n",line1);
   if(!strncmp(line1,"END",3)) goto label;
   j=2;
   /*     printf("\nIn the arc %s, number of points are",line1);*/
   fprintf(fpstat," %s,",line1);
   count=0;
   break;
 case 2:
   for(i=0;i<39 && (c=getc(fpin))!=EOF && c!= '\n';i++)
     line2[i]=c;
   line2[i]='\0';
   if(strncmp(line2,"END",3)){
     fprintf(fpout,"%s,%s,2\n",line2,line1);
     count++;
   }
   else {
     fprintf(fpout,"%s\n",line2);
     /*     printf(" %d  points\n",count);*/
     fprintf(fpstat,"%d\n",count);
     j=1;
   }
   break;
 }
```

```
 }
label:  printf("\nJob over\n\n");
  fclose(fpin);
  fclose(fpout);
  fclose(fpstat);
}
```

Appendix 4. A program written in the C-Programming language for automatically reading the lineament values including their length, angle and direction from the input vector sources such as ERDAS imagine or the ARC/INFO. Further, with the help of this program the following values are computed and prepared as an input for the computation using the Mathlab:

1. Output lineament statistics file,
2. Output stress statistics file,
3. Output raw data file for lineament number rose diagram,
4. Output raw data file for stress number rose diagram,
5. Output raw data file of length weighted lineament rose diagram and
6. Output raw data file for length weighted stress rose diagram.

```
/* Utility Program used by the main program */

#include<stdio.h>
#include<ctype.h>
#include<string.h>

main(){
 FILE *fparc,*fptemp;
 char c,line1[50],line2[50],line3[50],infile[20],outfile[20];
 int j=0;
 short i;
 printf("Give the input file name\t:");
 scanf("%s",infile);
 if(!(fparc=fopen(infile,"r")))
   {
    system("clear");
    printf("CANNOT OPEN FILENAME \"%s\" TO READ",infile);
    printf("\nCheck from the list below......\n\n");
    system("ls -la");
    printf("\nPROGRAM TERMINATED.. Re Execute\n\n\n\n");
    return(0);
   }
 printf("Give the output file name\t:");
 scanf("%s",outfile);
 for(i=0;i<40;i++){
   line2[i]=' ';
   line1[i]=' ';
 }
 for(i=0;i<39 && (c=getc(fparc))!=EOF && c!= '\n';i++)
   line1[i]=c;
 if(c=='\n'){
```

```
  line1[i]=c;
  i++;
}
line1[i]='\0';
for(i=0;i<39 && (c=getc(fparc))!=EOF && c!= '\n';i++)
  line2[i]=c;
if(c=='\n'){
  line2[i]=c;
  i++;
}
line2[i]='\0';
fptemp=fopen(outfile,"w");
while(!feof(fparc)){
  for(i=0;i<40;i++)
    line3[i]=' ';
  for(i=0;i<39 && (c=getc(fparc))!=EOF && c!= '\n';i++)
    line3[i]=c;
  if(c=='\n'){
    line3[i]=c;
    i++;
  }
  line3[i]='\0';
  if(c==EOF)
    break;
  /*   printf("Line 1 is %s\n",line1);
  printf("Line 2 is %s\n\n",line2);
  printf("Line 3 is %s\n\n",line3);
   getchar();
  */
  switch(j){
    case 0: fprintf(fptemp,"%s",line1);
```

```
        j=1;
        break;
   case 1: fprintf(fptemp,"%s",line1);
     j=2;
     break;
   case 2: if(strncmp(line3,"END",3)==0){
     fprintf(fptemp,"%s",line1);
     fprintf(fptemp,"%s",line3);
     j=3;
   }
     break;
   case 3:
     j=4;
     break;
   case 4:
     j=0;
     break;
   }
   strncpy(line1,line2,40);
   strncpy(line2,line3,40);
  }
     fclose(fparc);
  fclose(fptemp);
}
/*  Main program*/
#include<stdio.h>
#include<math.h>
#include<ctype.h>
#include<string.h>
#define PI 3.1415927
#define BINWIDTH 10
```

```
main(){
 FILE *fparc,*fptemp;
 FILE *fpoutlin, *fpoutstress;
 FILE *fprawlin_no, *fprawstress_no;
 FILE *fprawlin_le, *fprawstress_le;
 char c,line[50],infile[20];
 short i,j;
 float theta,alpha,phi,beta1,beta2;
 float thetasumlength[360/BINWIDTH],betasumlength[360/BINWIDTH];
 int k,thetanolines[360/BINWIDTH], betanolines[360/BINWIDTH];
 float length,x1,x2,y1,y2;
 system("clear");
 printf("\n\n\nGive the input file name(ERDAS, R2V, ARC INFO)\t:");
 scanf("%s",infile);
 if(!(fparc=fopen(infile,"r")))
  {
   system("clear");
   printf("CANNOT OPEN FILENAME \"%s\" TO READ",infile);
   printf("\nCheck from the list below......\n\n");
   system("ls -la");
   printf("\nPROGRAM TERMINATED.. Re Execute\n\n\n\n");
   return(0);
  }
 printf("Enter the value of PHI in degrees(eg. 30.0)\t:");
 scanf("%f",&phi);
 printf("\n\nJob state.........Working");
 phi *= PI/180.0;
 alpha = (PI/4.0) - (phi/2.0);
 fptemp=fopen("temp.dat","w");
 j=1;
```

```
while(!feof(fparc)){
 for(i=0;i<40;i++)
  line[i]=' ';
 for(i=0;i<39 && (c=getc(fparc))!=EOF && c!= '\n';i++)
  if(c!=',') line[i]=c;
 if(c=='\n'){
  line[i]=c;
  i++;
 }
 if(c==EOF){
  line[i]='\0';
  if(((j%4)!=1) && ((j%4)!=0))
    fprintf(fptemp,"%s",line);
  break;
 }
  line[i]='\0';
  if(((j%4)!=1) && ((j%4)!=0))
   fprintf(fptemp,"%s",line);
 j++;
}
fclose(fparc);
fclose(fptemp);
getchar();
fptemp=fopen("temp.dat","r");
for(k=0;k<180/BINWIDTH;k++){
 thetasumlength[k]=0.0;
 betasumlength[k]=0.0;
 thetanolines[k]=0;
 betanolines[k]=0;
}
while(!feof(fptemp)){
```

```
  fscanf(fptemp,"%f %f %f %f",&x1,&y1,&x2,&y2);
  if(feof(fptemp))
   break;
  theta=atan((y2-y1)/(x2-x1));
  if(theta<0.0)
   theta=theta+PI;
  length=sqrt((x1-x2)*(x1-x2) + (y1-y2)*(y1-y2));
  thetasumlength[(int)(theta*180.0/(BINWIDTH*PI))]+=length;
  thetanolines[(int)(theta*180.0/(BINWIDTH*PI))]++;
  beta1=theta + alpha;
  if(beta1>PI)
   beta1-=PI;
  betasumlength[(int)(beta1*180.0/(BINWIDTH*PI))]+=length;
  betanolines[(int)(beta1*180.0/(BINWIDTH*PI))]++;
  beta2=theta - alpha;
  if(beta2<0.0)
   beta2 += PI;
  betasumlength[(int)(beta2*180.0/(BINWIDTH*PI))]++;
  betanolines[(int)(beta2*180.0/(BINWIDTH*PI))]++;
 }
 fpoutlin=fopen("Stat_Lin.dat","w");
 fpoutstress=fopen("Stat_Stress.dat","w");
 fprawlin_no= fopen("Raw_Lin_Num.dat","w");
 fprawstress_no= fopen("Raw_Stress_Num.dat","w");
 fprawlin_le= fopen("Raw_Lin_Len.dat","w");
 fprawstress_le= fopen("Raw_Stress_Len.dat","w");
 for(k=0;k<180.0/BINWIDTH;k++){
fprintf(fpoutlin,"%f\t%f\t%f\t\t%d\n",k*(float)BINWIDTH,(k+1)*(float)BINWIDTH,theta
sumlength[k],thetanolines[k]);
fprintf(fpoutstress,"%f\t%f\t%f\t\t%d\n",k*(float)BINWIDTH,(k+1)*(float)BINWIDTH,b
etasumlength[k],betanolines[k]);
 for(i=0;i<thetanolines[k];i++)
```

```
        fprintf(fprawlin_no,"%f\n",(k+0.5)*(float)BINWIDTH);
for(i=0;i<betanolines[k];i++)
        fprintf(fprawstress_no,"%f\n",(k+0.5)*(float)BINWIDTH);
for(i=0;i<(int)((thetasumlength[k]+500.0)/1000.0);i++)
        fprintf(fprawlin_le,"%f\n",(k+0.5)*(float)BINWIDTH);
for(i=0;i<(int)((betasumlength[k]+500.0)/1000.0);i++)
        fprintf(fprawstress_le,"%f\n",(k+0.5)*(float)BINWIDTH);
}
fclose(fpoutlin);
fclose(fpoutstress);
fclose(fprawlin_le);
fclose(fprawlin_no);
fclose(fprawstress_le);
fclose(fprawstress_no);
fclose(fptemp);
system("rm -f temp.dat");
printf("\n\n\nOutput lineament statistics file is Stat_Lin.dat.\n");
printf("\nOutput stress statistics file is Stat_Stress.dat.\n");
printf("\nOutput raw data file for lineament number rose diagram is
Raw_Lin_Num.dat\n");
printf("\nOutput raw data file for stress number rose diagram is Raw_Stress_Num.dat\n");
printf("\nOutput raw data file of lengthweighted lineament rose diagram is
Raw_Lin_Len.dat\n");
printf("\nOutput raw data file for lengthweighted stress rose diagram is
Raw_Stress.dat\n\n");
printf("\n\n");
}
```

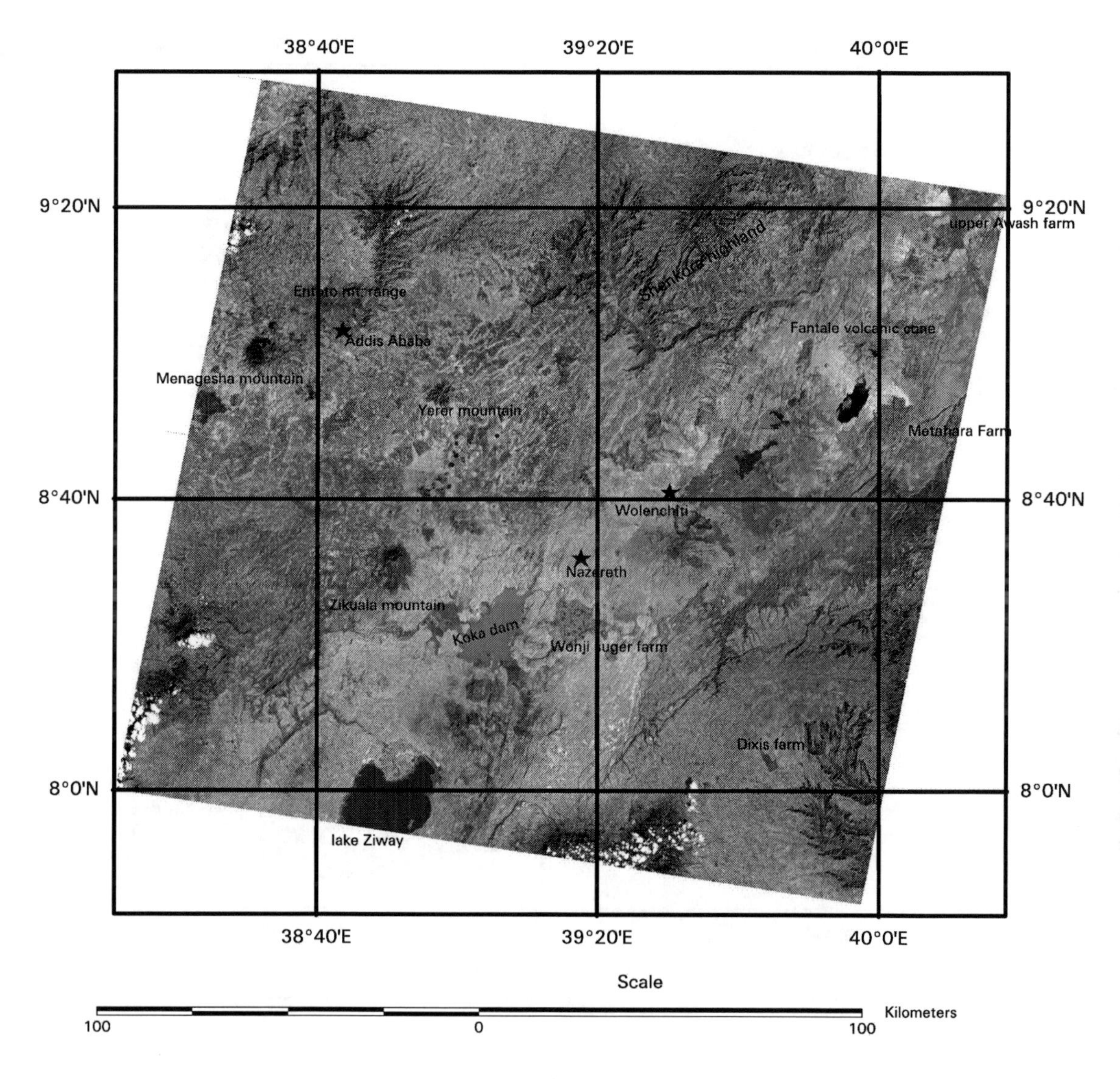

Map 1. Landsat TM full scene after transforming in IHS, stretchning of the saturation and back transformation to RGB. Band combination (4,7,5) in RGB. Please notice the contrast difference of the four merged sub-scenes which was caused due to pre-delivery correction at the EOSAT.

by Mezemir Fikre-Mariam Wagaw, Feb. 2001
Institute of Geography
Faculty VII - Architecture Environment and Society
Technical University of Berlin

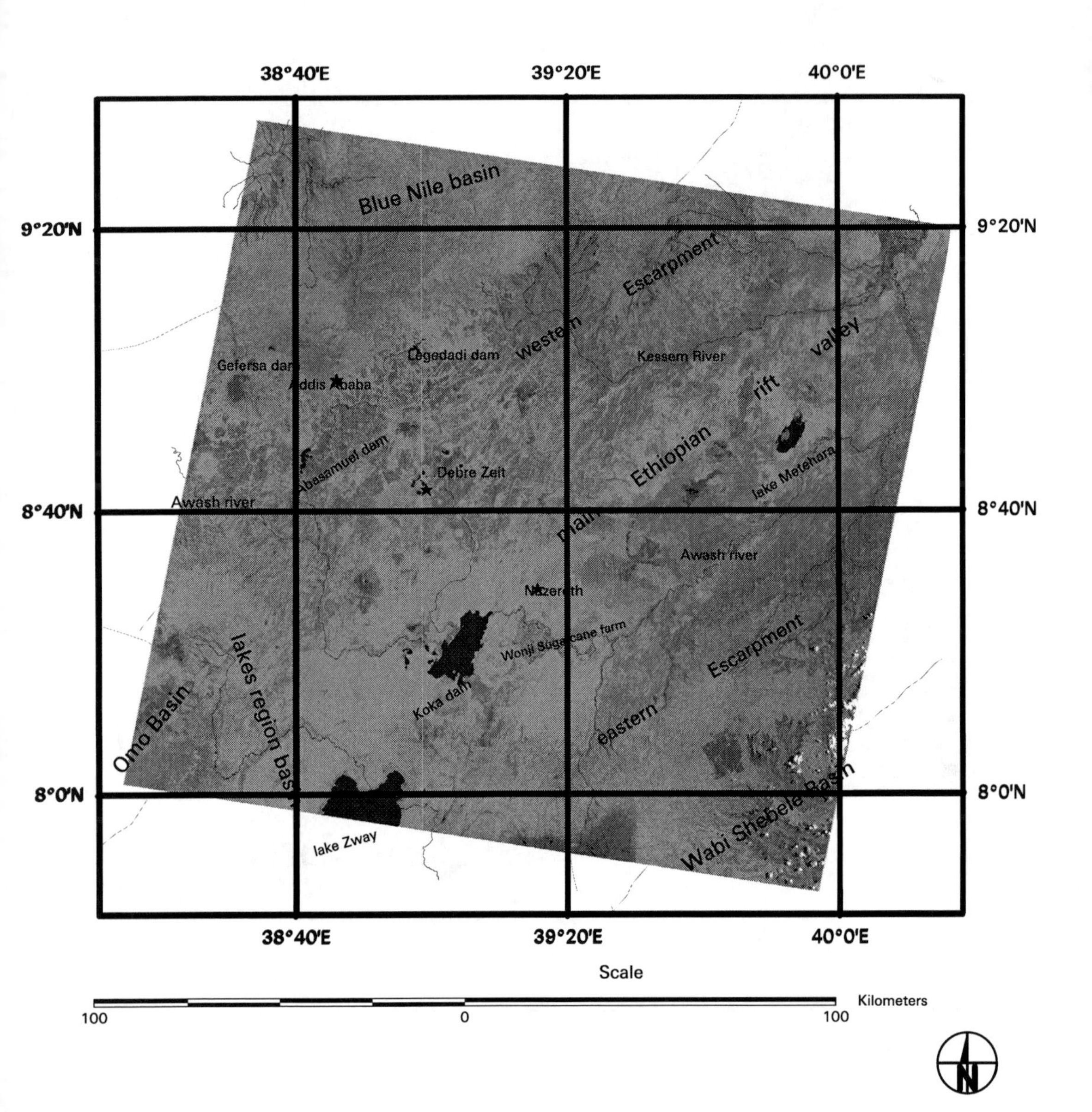

Map 2. Maximum likelihood classification applied on the PCI transformed Landsat MSS, overlaid with the main rivers and water shade coverage. The blue class represents the surface water body of the area. In this classification the cloud shade - mainly at the south east area - and the water body are resulted in to the same pixel class.

by Mezemir Fikre-Mariam Wagaw, Feb. 2001
Institute of Geography
Faculty VII - Architecture Environment and Society
Technical University Berlin

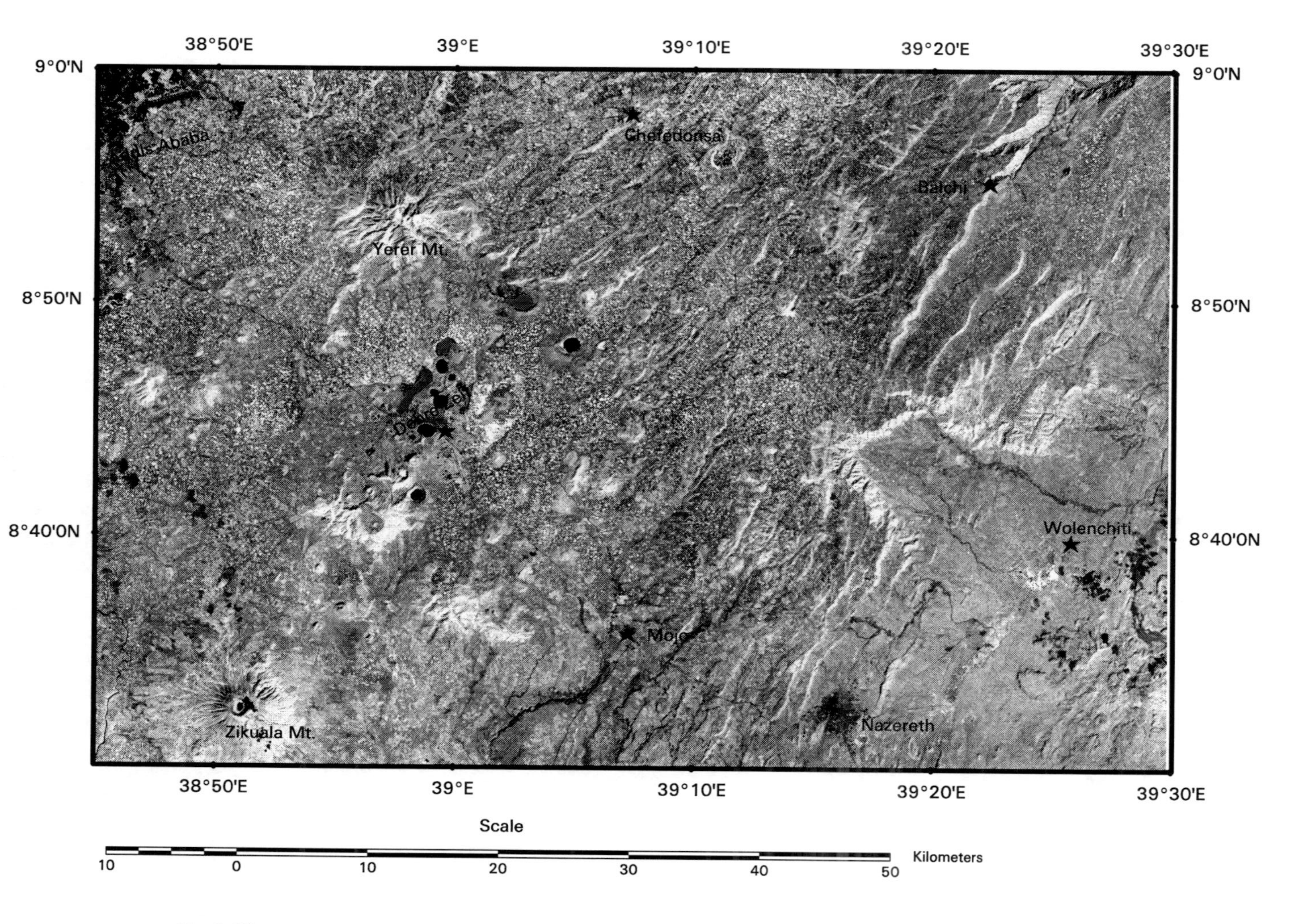

Map3. Histogram equalizeed Landsat TM Principal Component color composite (4,3,2) in RGB with better background (soil/rock) and agricultural field information.

by Mezemir Fikre-Mariam Wagaw, Feb. 2001
Institute of Geography
Faculty VII - Architecture Environment and Society
Technical University of Berlin

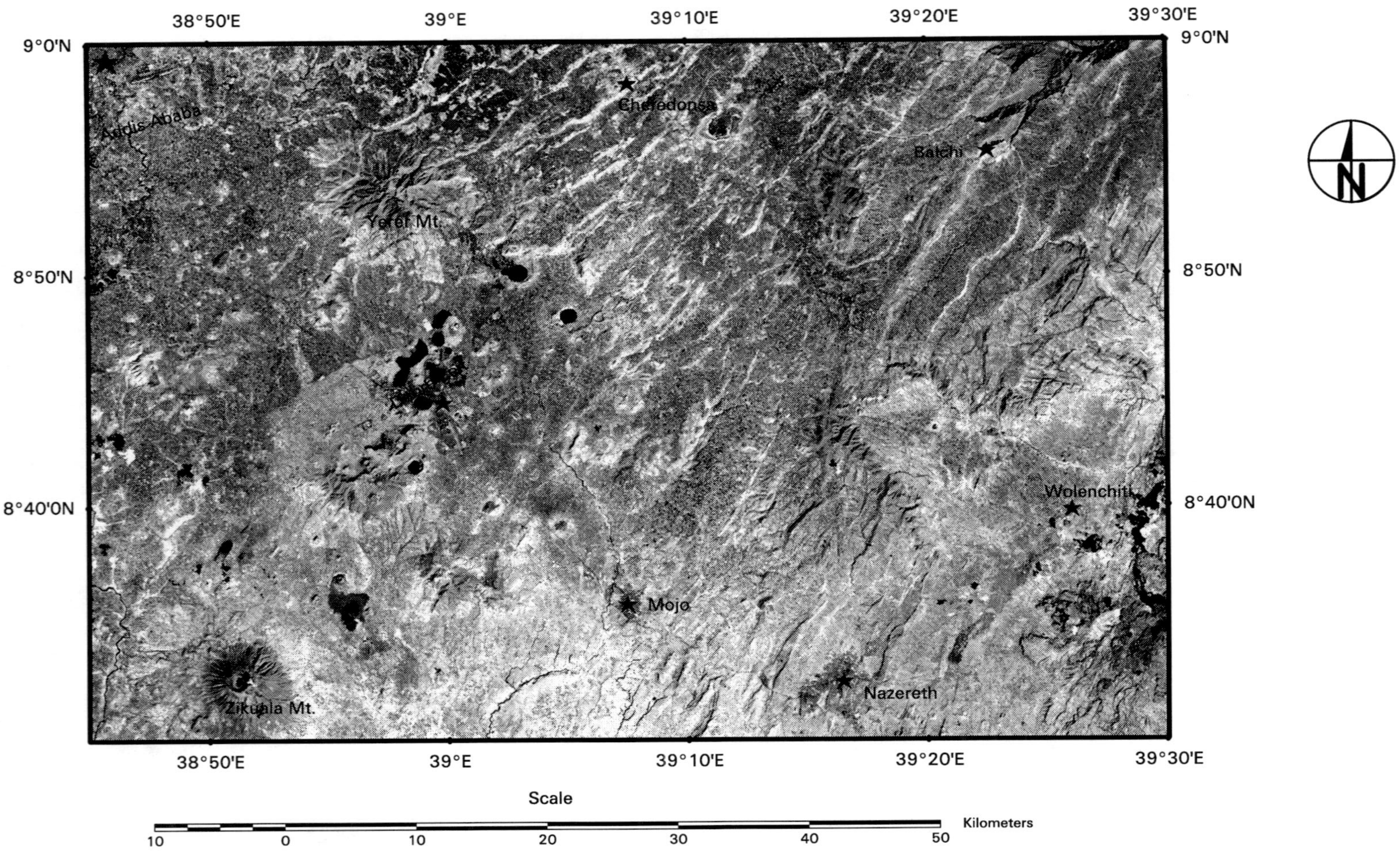

Map 4. Histogram equalizeed Landsat TM Principal Component color composite (3,1,2) in RGB with better geology and geomorphology contrast.

by Mezemir Fikre-Mariam Wagaw, Feb. 2001
Institute of Geography
Faculty VII - Architecture Environment and Society
Technical University of Berlin

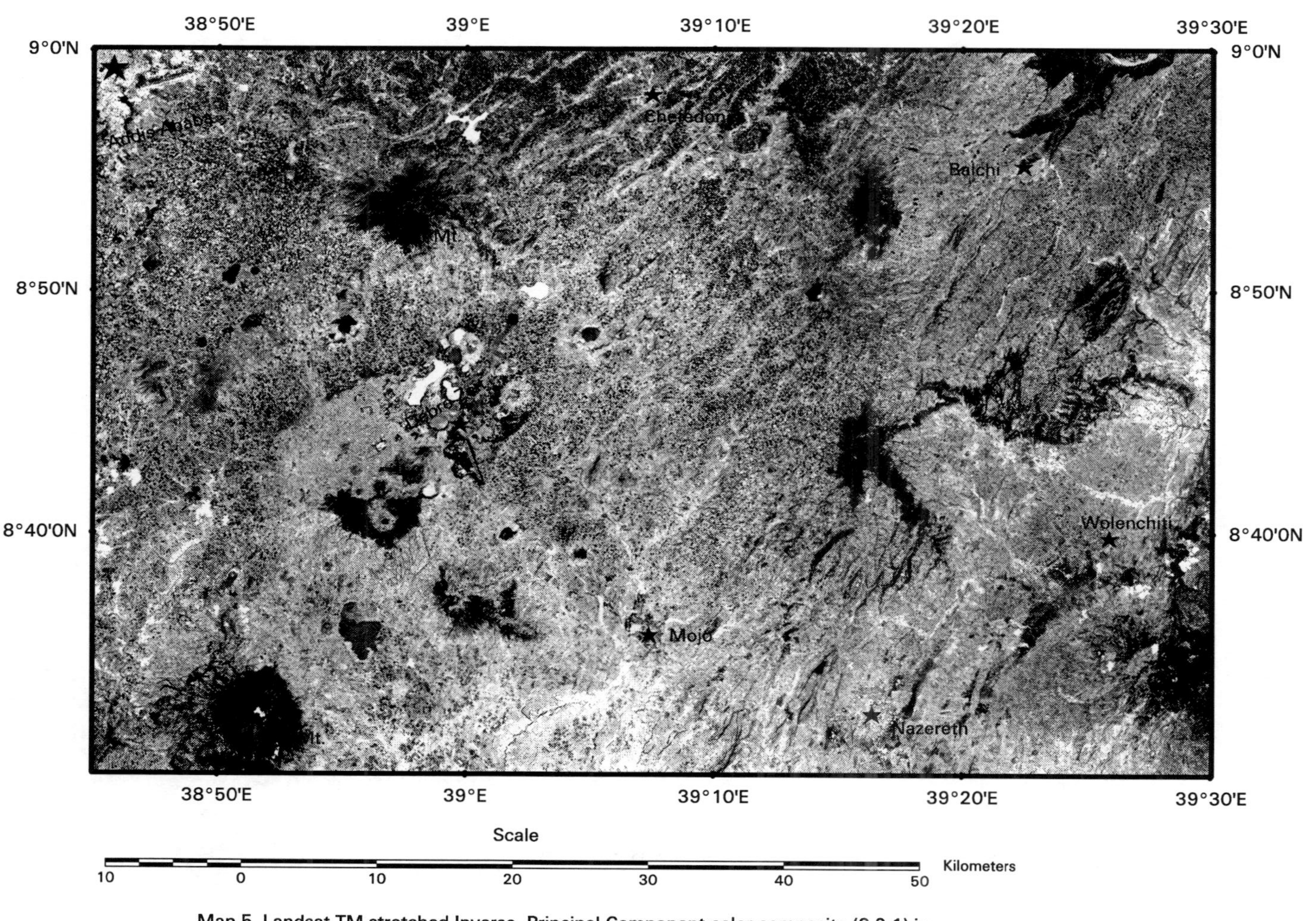

Map 5. Landsat TM stretched Inverse Principal Component color composite (3,2,1) in RGB with better contrast of the recent volcanic centers and linear structures.

by Mezemir Fikre-Mariam Wagaw, Feb. 2001
Institute of Geography
Faculty VII - Architecture Environment and Society
Technical University of Berlin

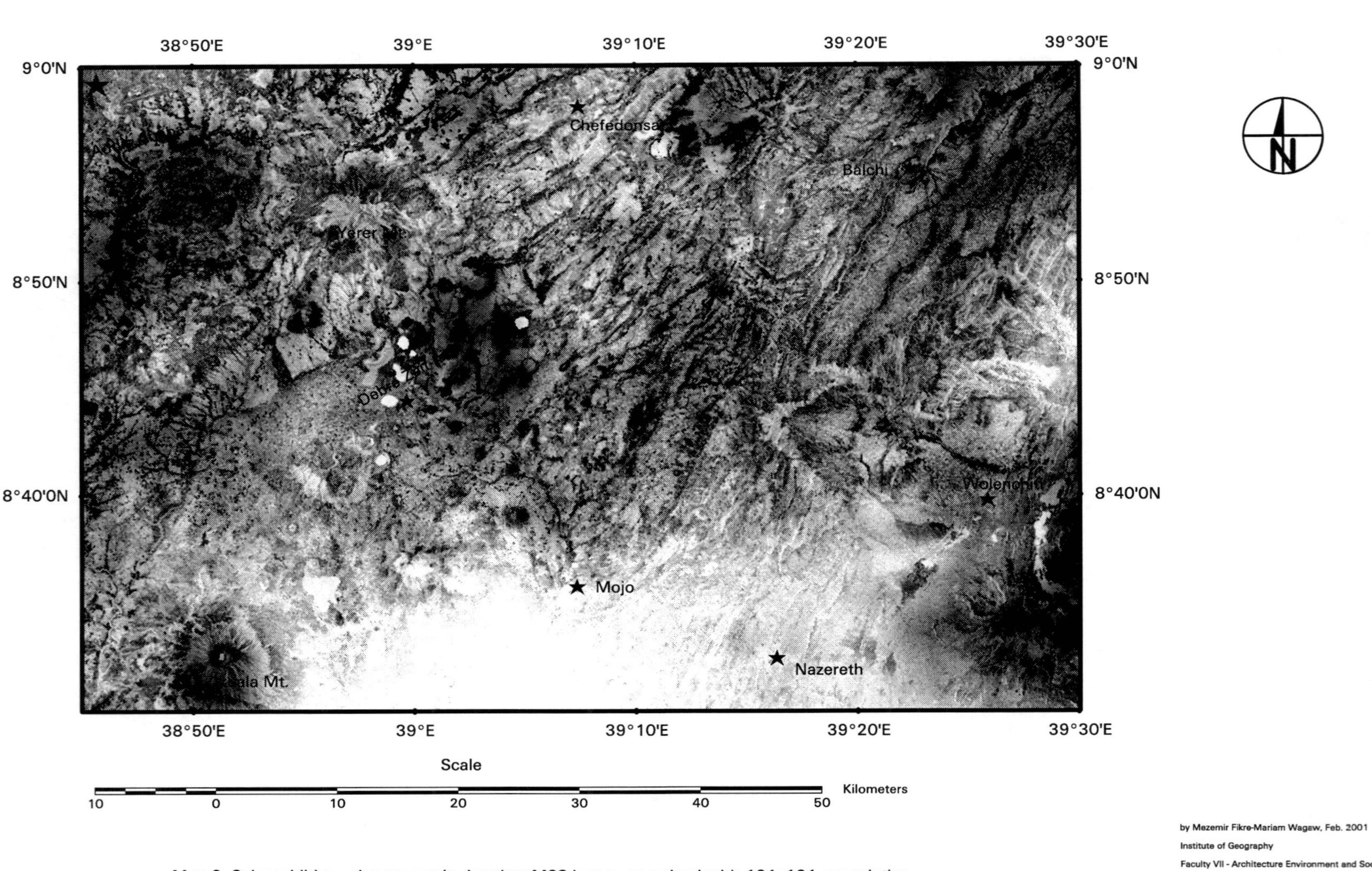

Map 6. Color additive color composite Landsat MSS image convolved with 101x101 convolution matrix, band combination (4,3,1) in RGB.

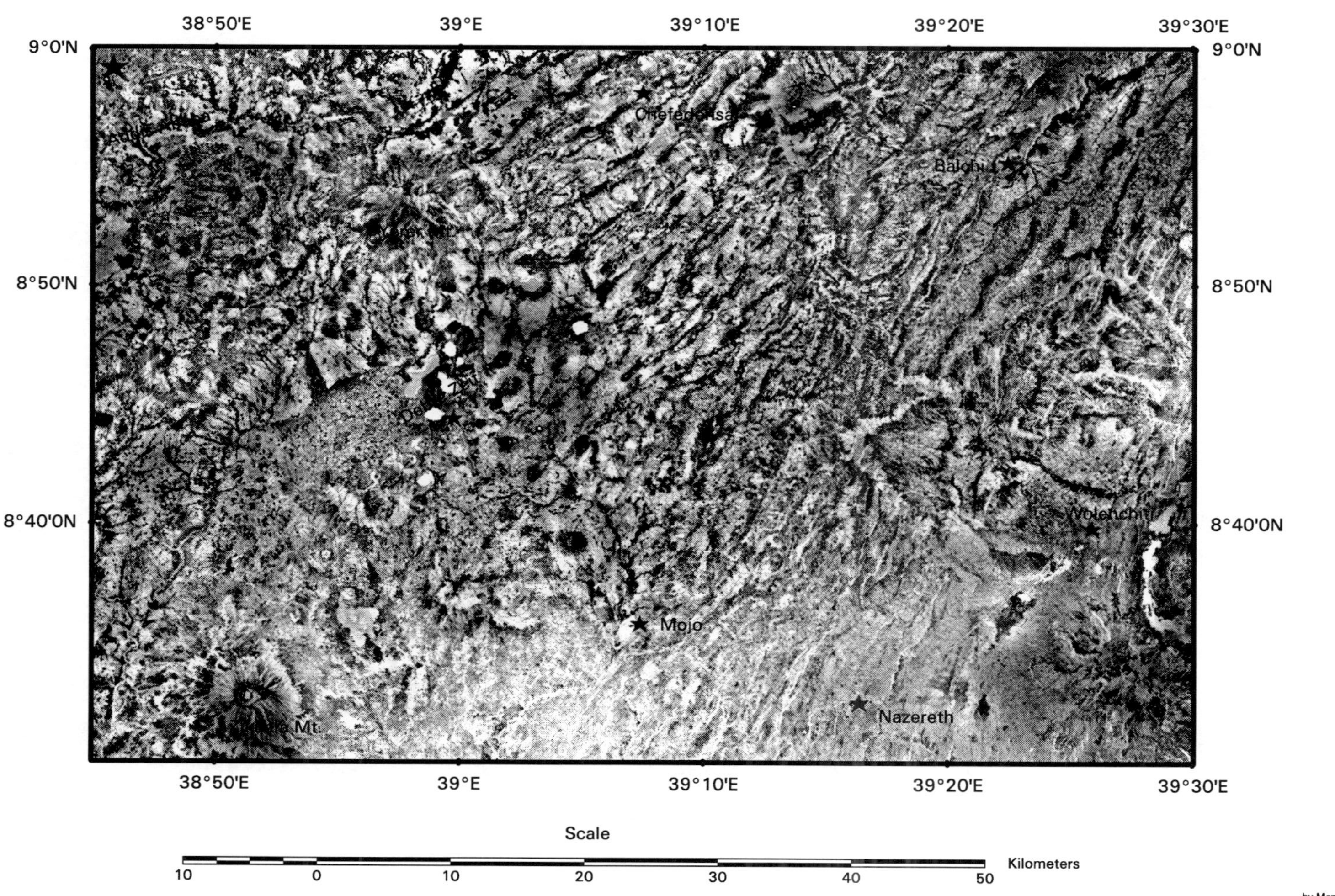

Map 7. Color additive color composite Landsat MSS image convolved with 51x51 convolution matrix, band combination (4,2,1) in RGB.

by Mezemir Fikre-Mariam Wagaw, Feb. 2001
Institute of Geography
Faculty VII - Architecture Environment and Society
Technical University of Berlin

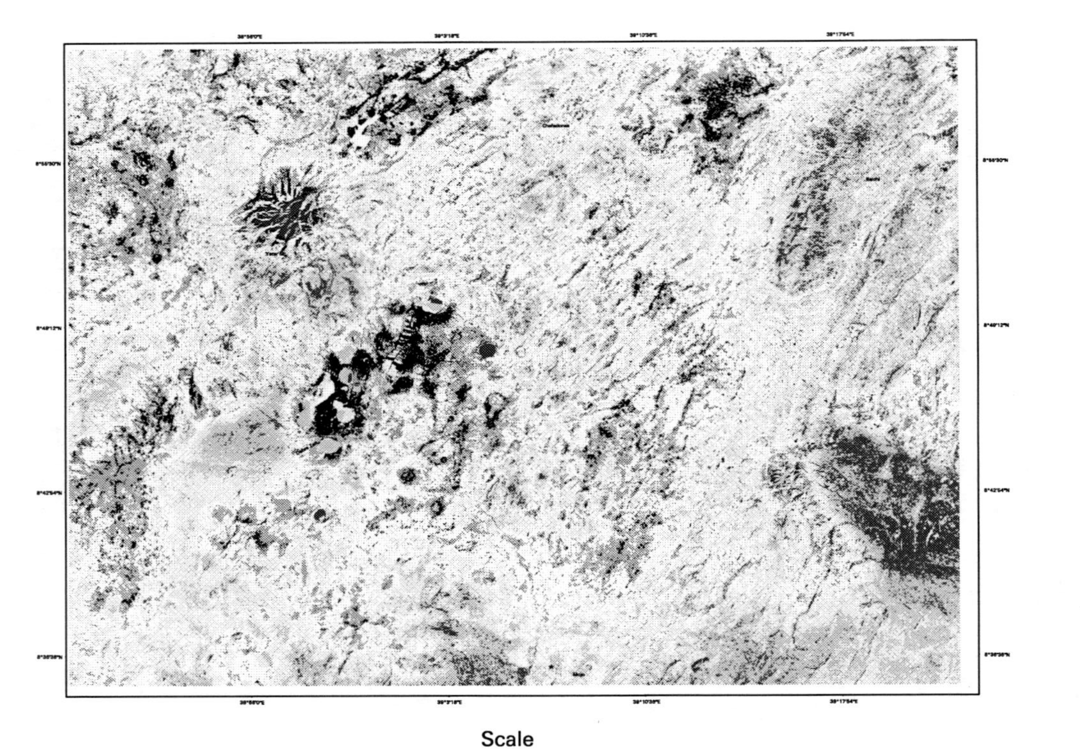

Map 8. Landsat TM with a primary convolution using a 101x101 convolution matrix and followed by an unsupervised classification into 6 classes.

Legend for the unsupervised classification classes

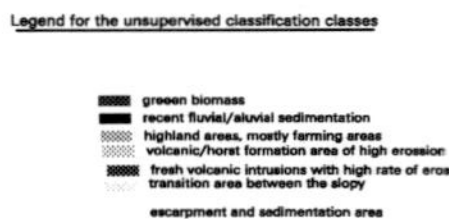

by Mezemir Fikre-Mariam Wagaw, Feb. 2001
Institute of Geography
Faculty VII - Architecture Environment and Society
Technical University of Berlin

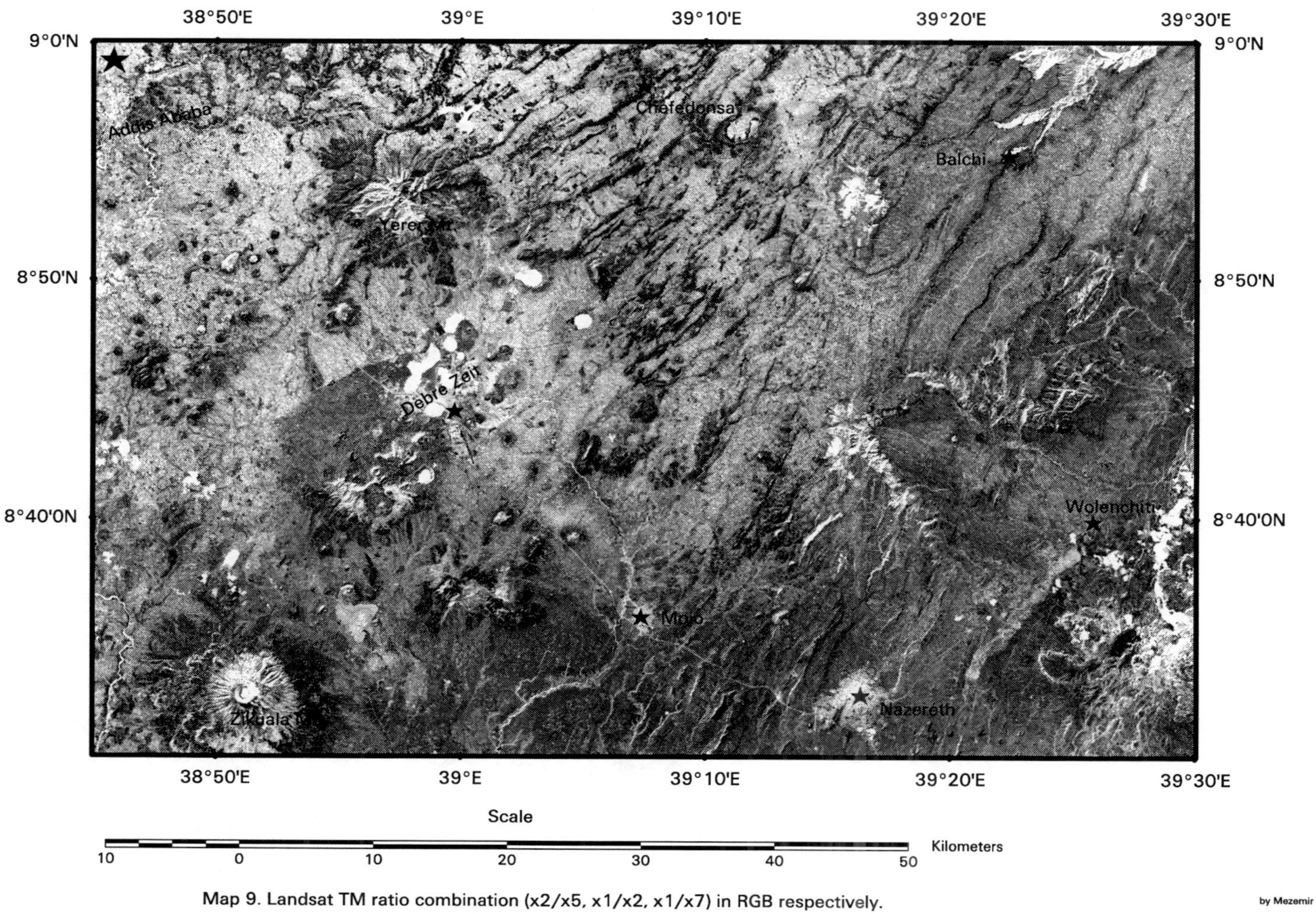

Map 9. Landsat TM ratio combination (x2/x5, x1/x2, x1/x7) in RGB respectively.

This processing enhanced the morphology and structural setup information of the area.

by Mezemir Fikre-Mariam Wagaw, Feb. 2001
Institute of Geography
Faculty VII - Architecture Environment and Society
Technical University of Berlin

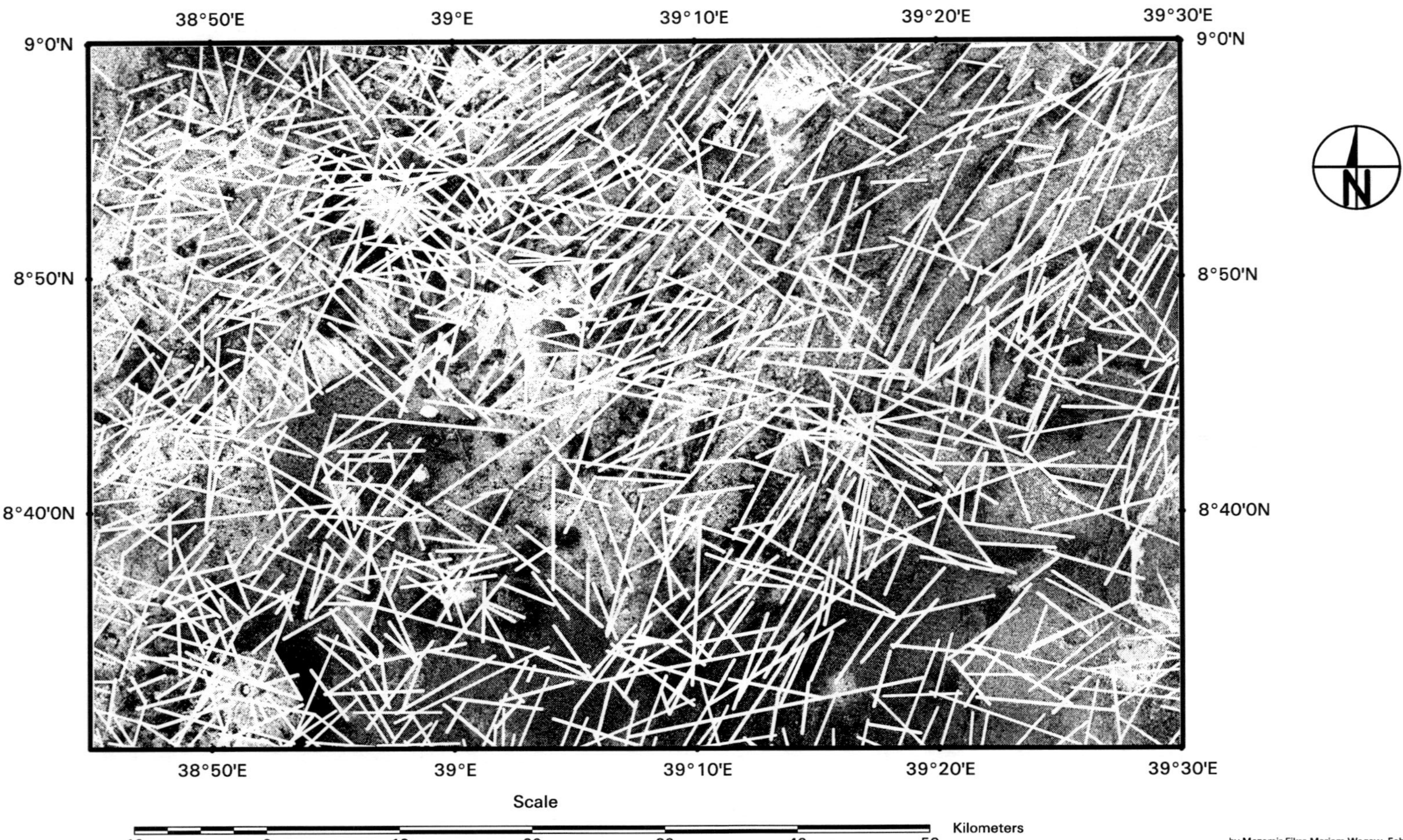

Map 10. Color additive color composite Landsat MSS ratio image (4/7, 4/5, 4/6) in RGB overlaid with the main vectorized lineaments.

by Mezemir Fikre-Mariam Wagaw, Feb. 2001
Institute of Geography
Faculty VII - Architecture Environment and Society
Technical University of Berlin

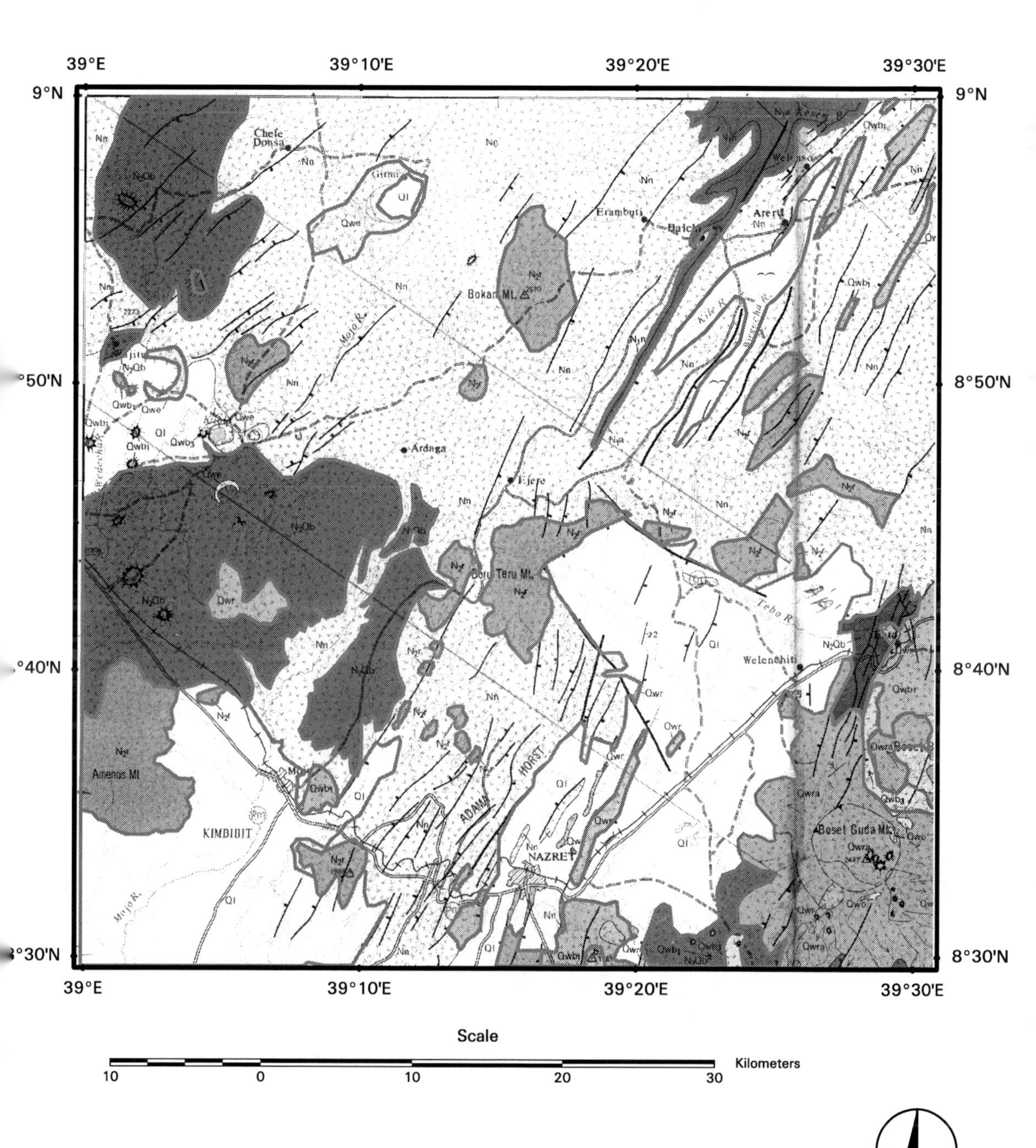

Map 11. Geological map of the study area, a cutout from the geology of Nazereth with the EIGS map (of the year 1978) original scale 1:250000.

by Mezemir Fikre-Mariam Wagaw, Feb. 2001
Institute of Geography
Faculty VII - Architecture Environment and Society
Technical University of Berlin

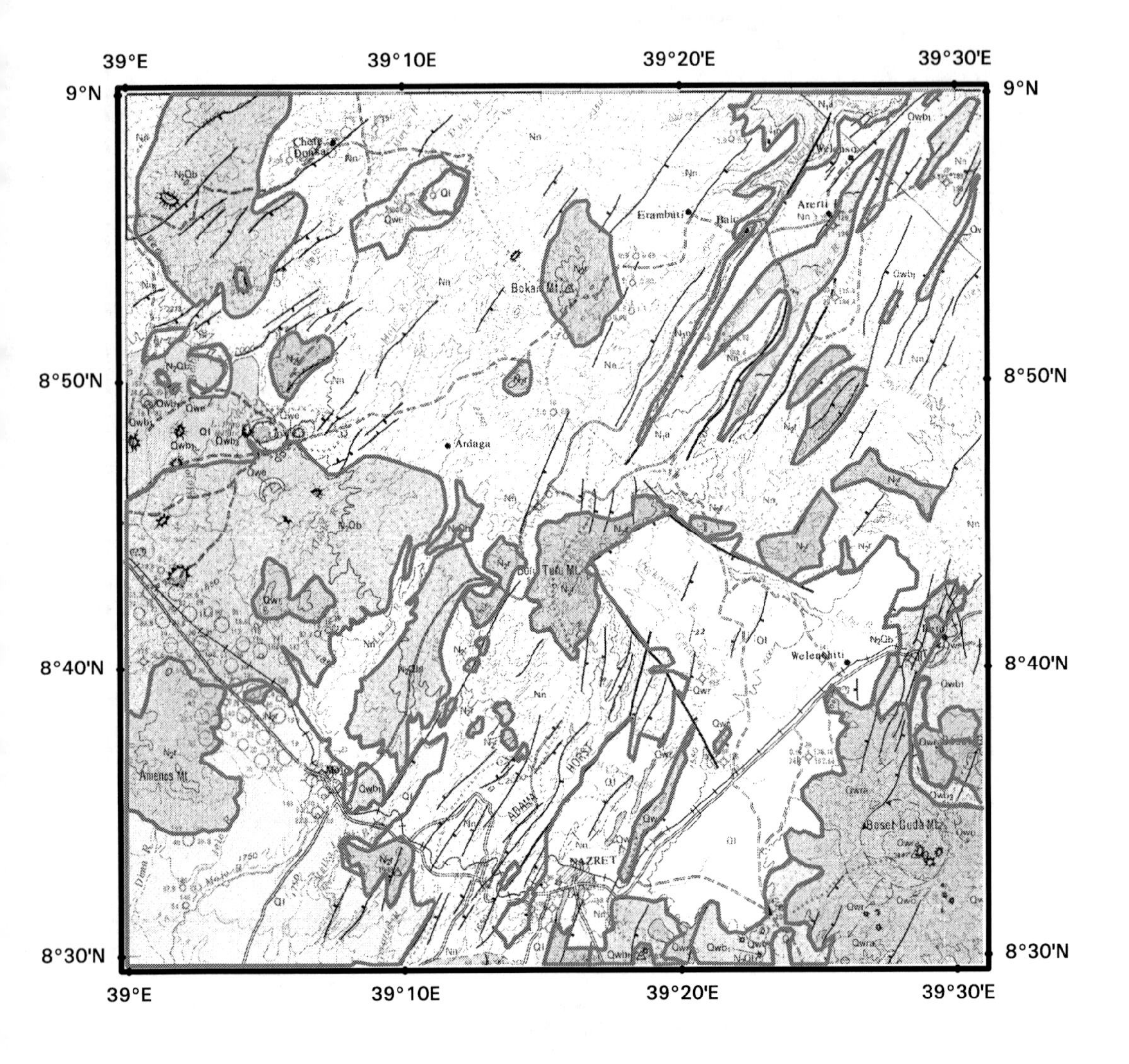

Scale

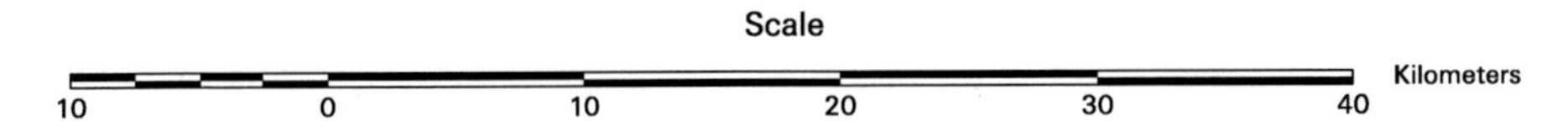

Map 12. Hydrogeological map of the study area, a cutout from the hydrogeological map of Nazereth, EIGS original scale 1:250000.

by Mezemir Fikre-Mariam Wagaw, Feb. 2001
Institute of Geography
Faculty VII - Architecture Environment and Society
Technical University of Berlin

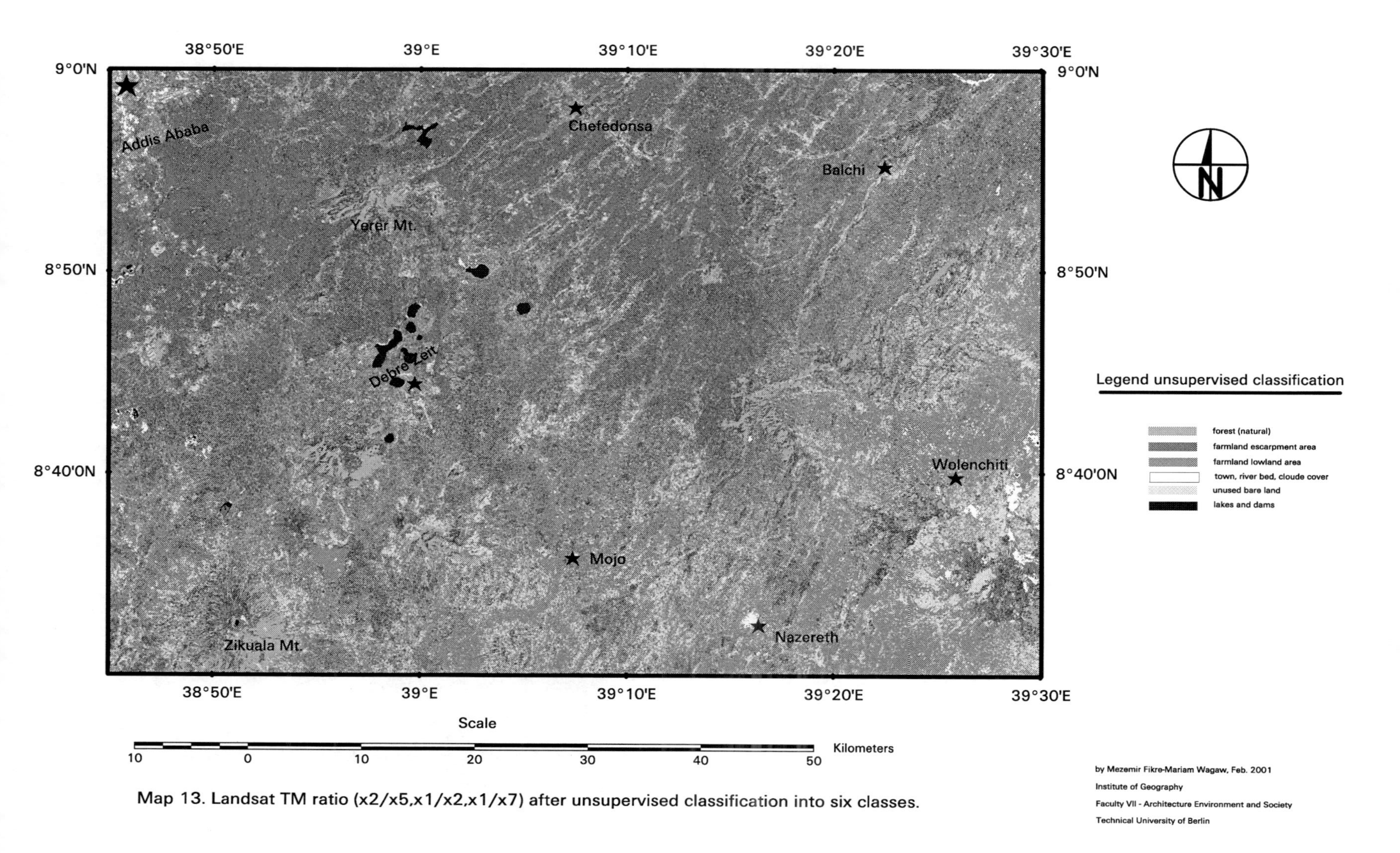

Map 13. Landsat TM ratio (x2/x5,x1/x2,x1/x7) after unsupervised classification into six classes.

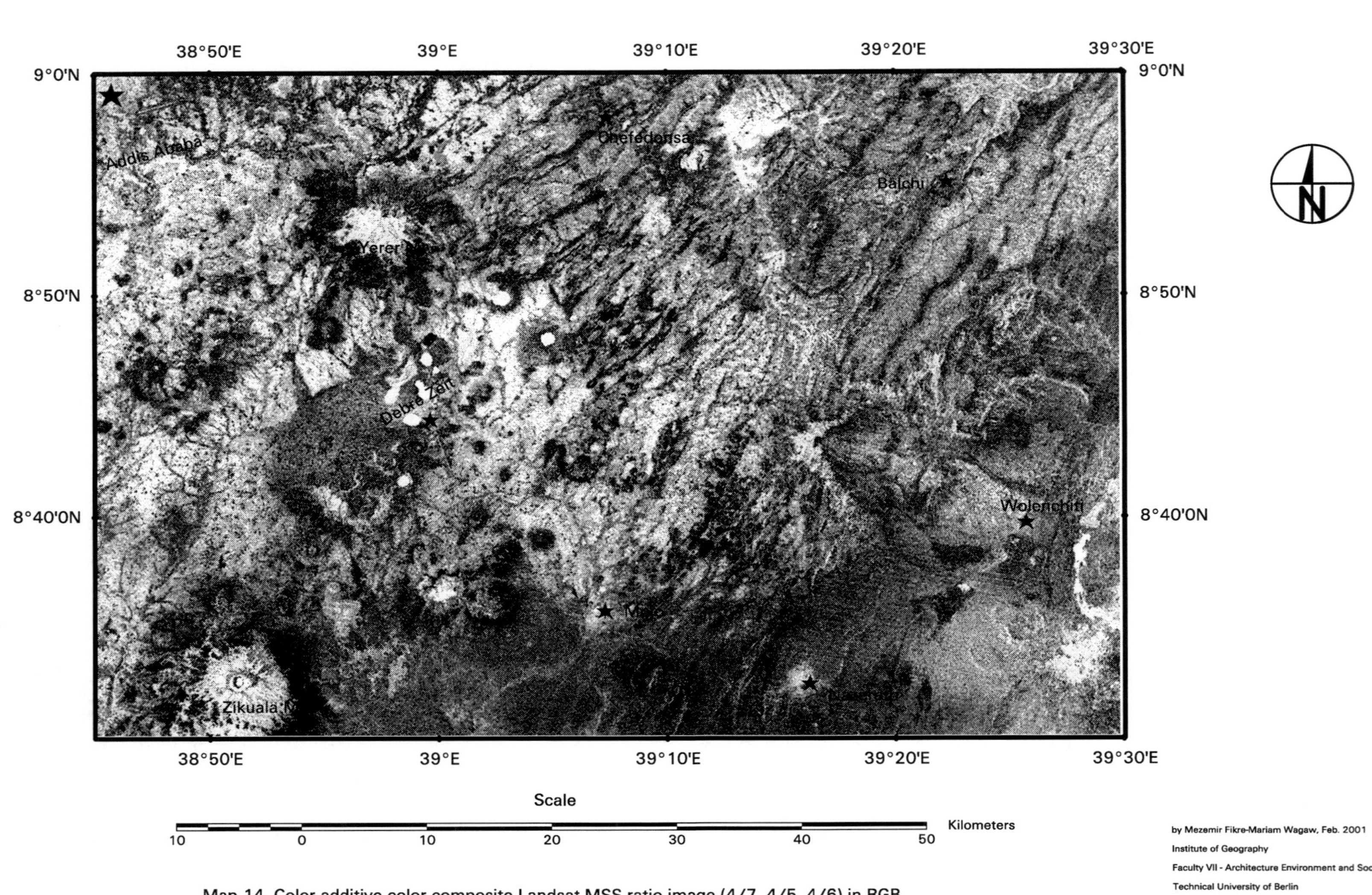

Map 14. Color additive color composite Landsat MSS ratio image (4/7, 4/5, 4/6) in RGB.

by Mezemir Fikre-Mariam Wagaw, Feb. 2001
Institute of Geography
Faculty VII - Architecture Environment and Society
Technical University of Berlin

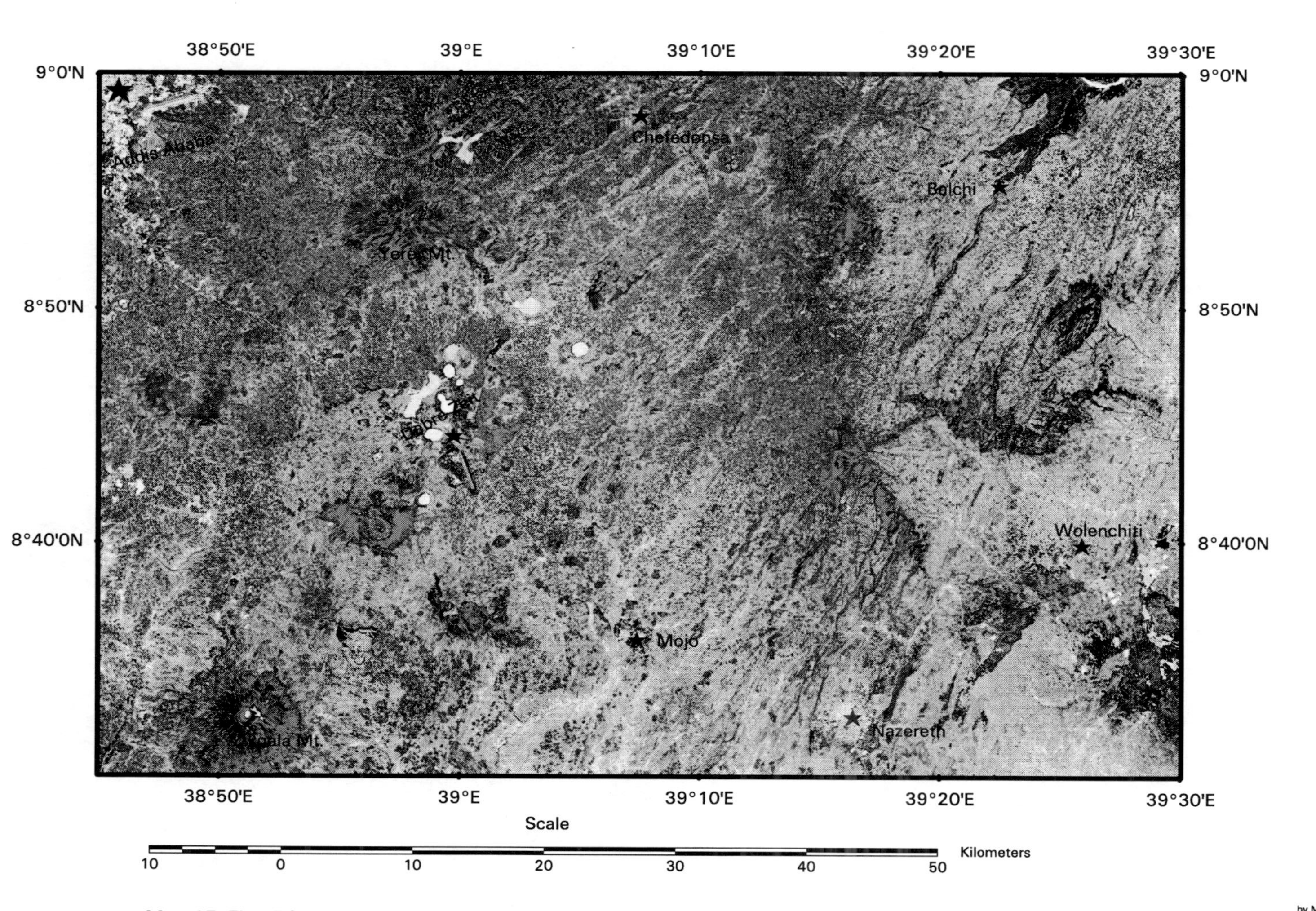

Map 15. First PC transformation and back transformation after stretching in the PC domain. The resulting components (4,3,2) were then transformed into IHS domain, stretched and back transformed into RGB.

by Mezemir Fikre-Mariam Wagaw, Feb. 2001
Institute of Geography
Faculty VII - Architecture Environment and Society
Technical University of Berlin

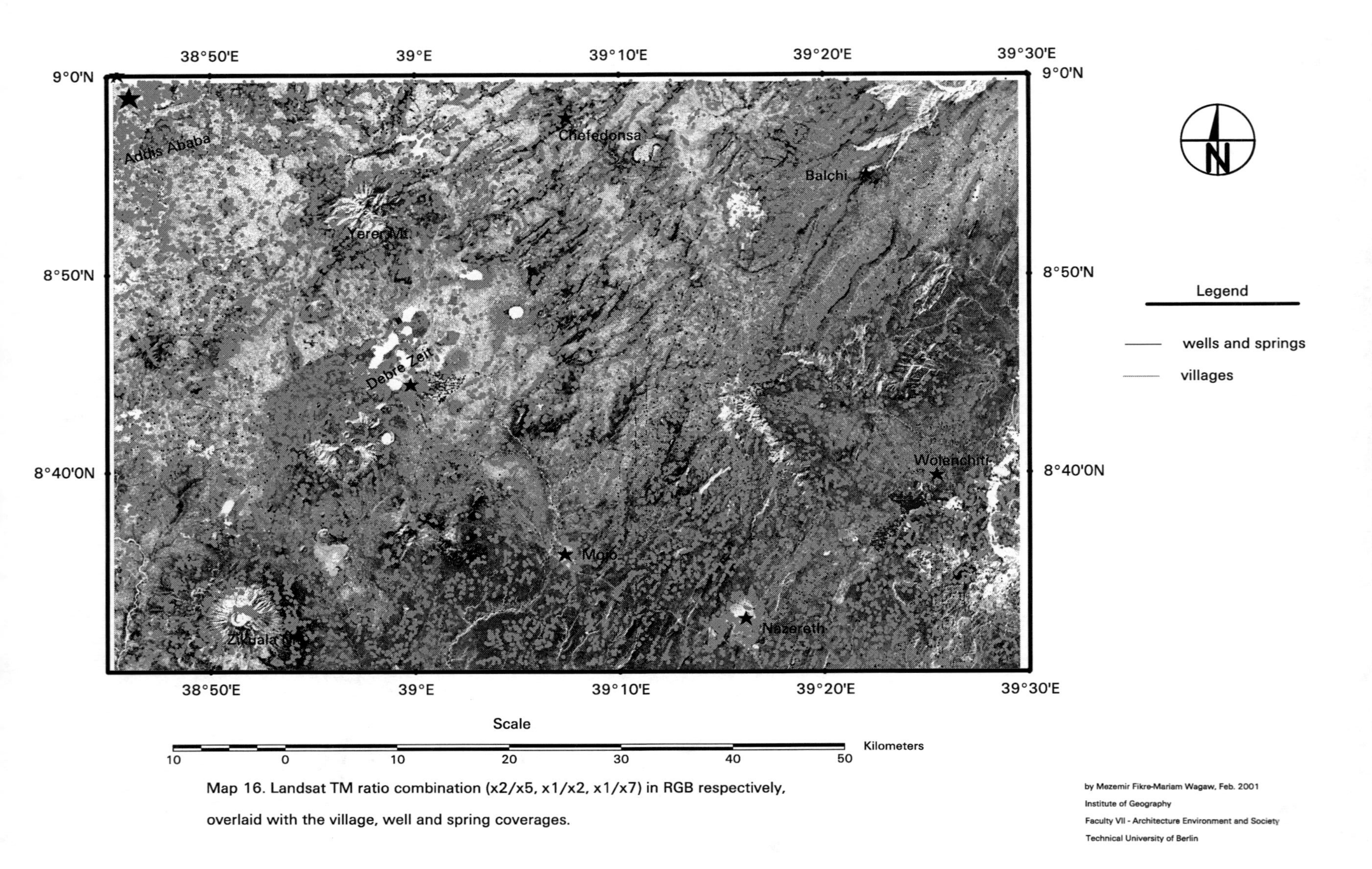

Map 16. Landsat TM ratio combination (x2/x5, x1/x2, x1/x7) in RGB respectively, overlaid with the village, well and spring coverages.

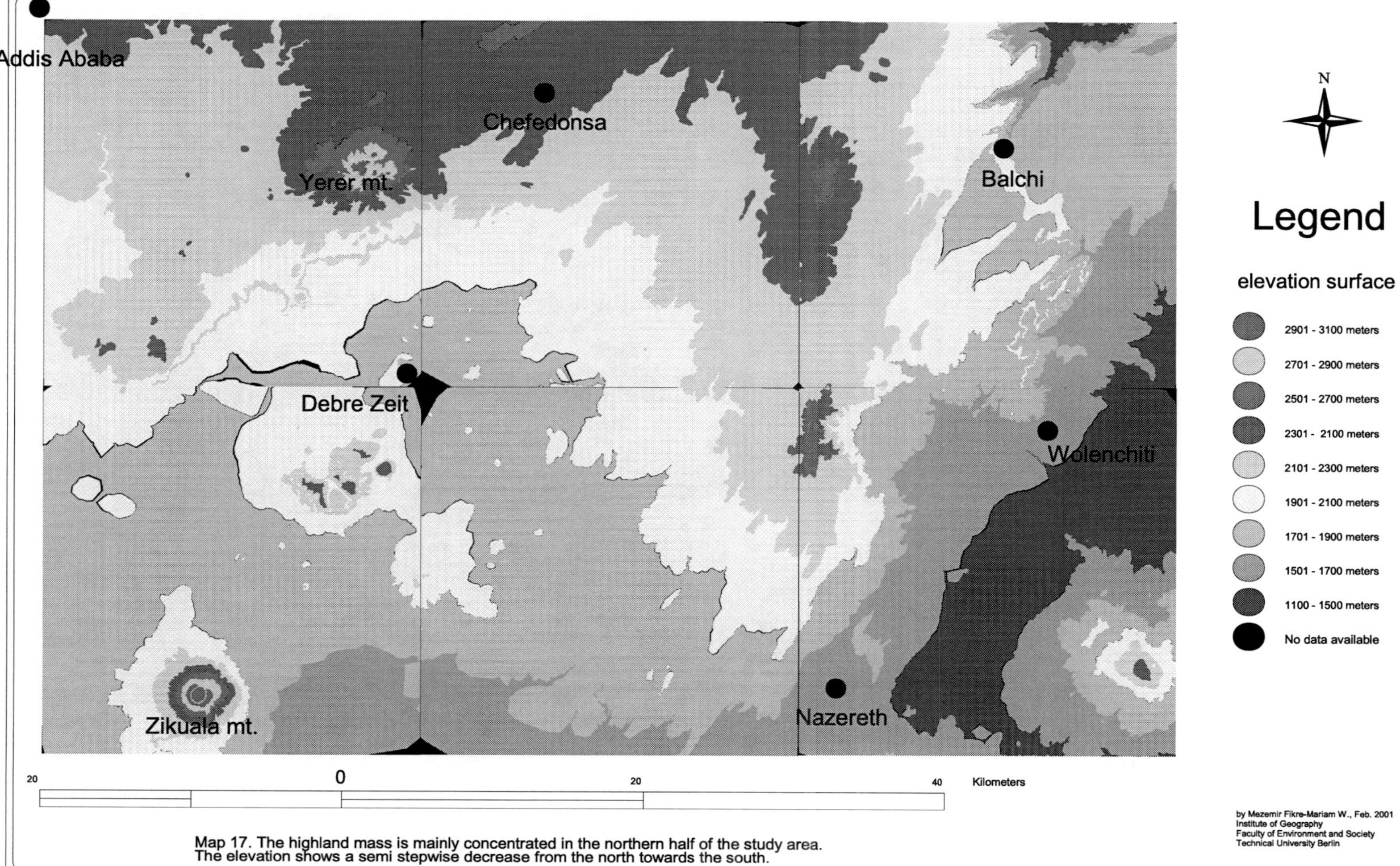

Map 17. The highland mass is mainly concentrated in the northern half of the study area. The elevation shows a semi stepwise decrease from the north towards the south.

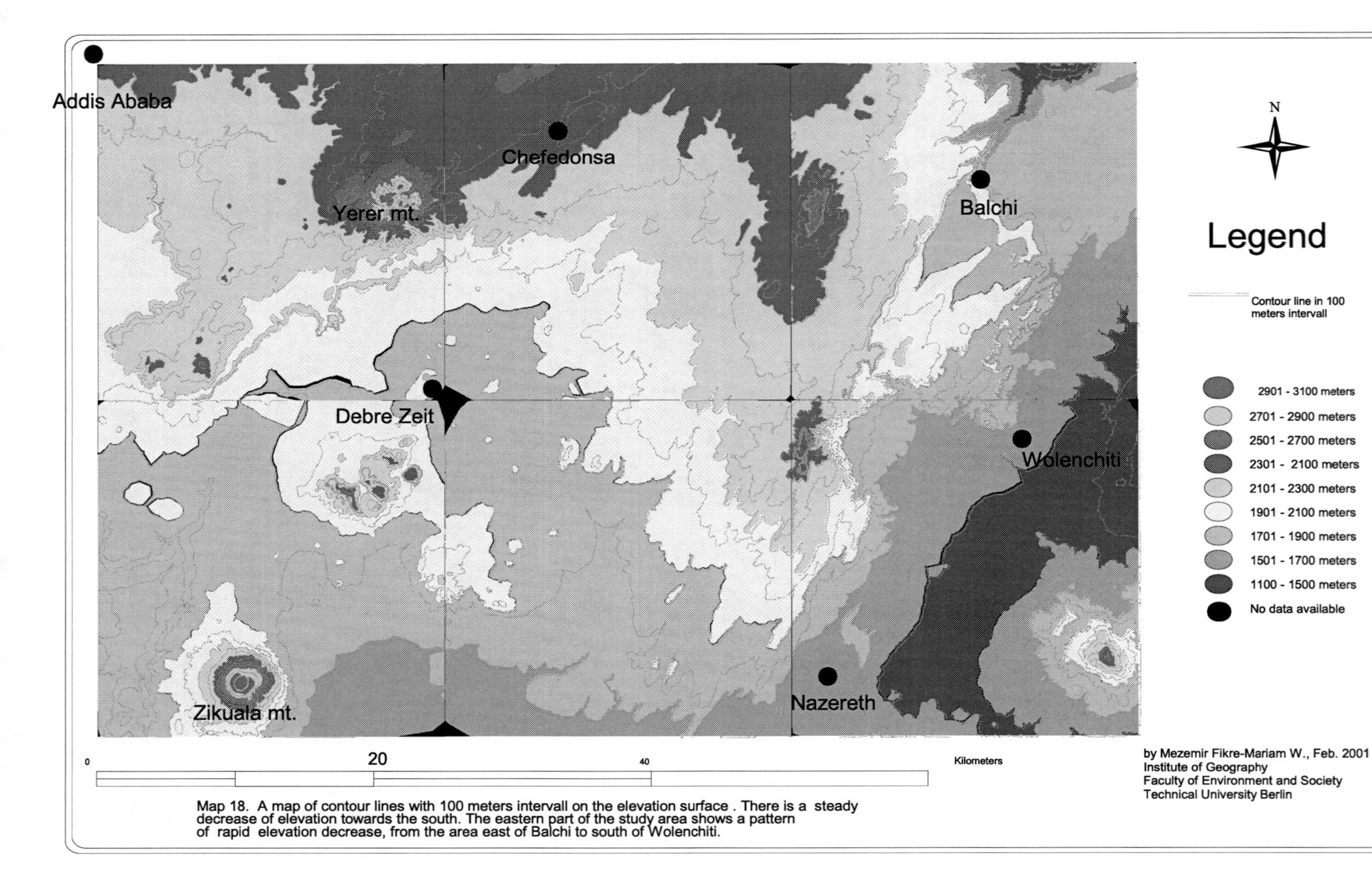

Map 18. A map of contour lines with 100 meters intervall on the elevation surface . There is a steady decrease of elevation towards the south. The eastern part of the study area shows a pattern of rapid elevation decrease, from the area east of Balchi to south of Wolenchiti.

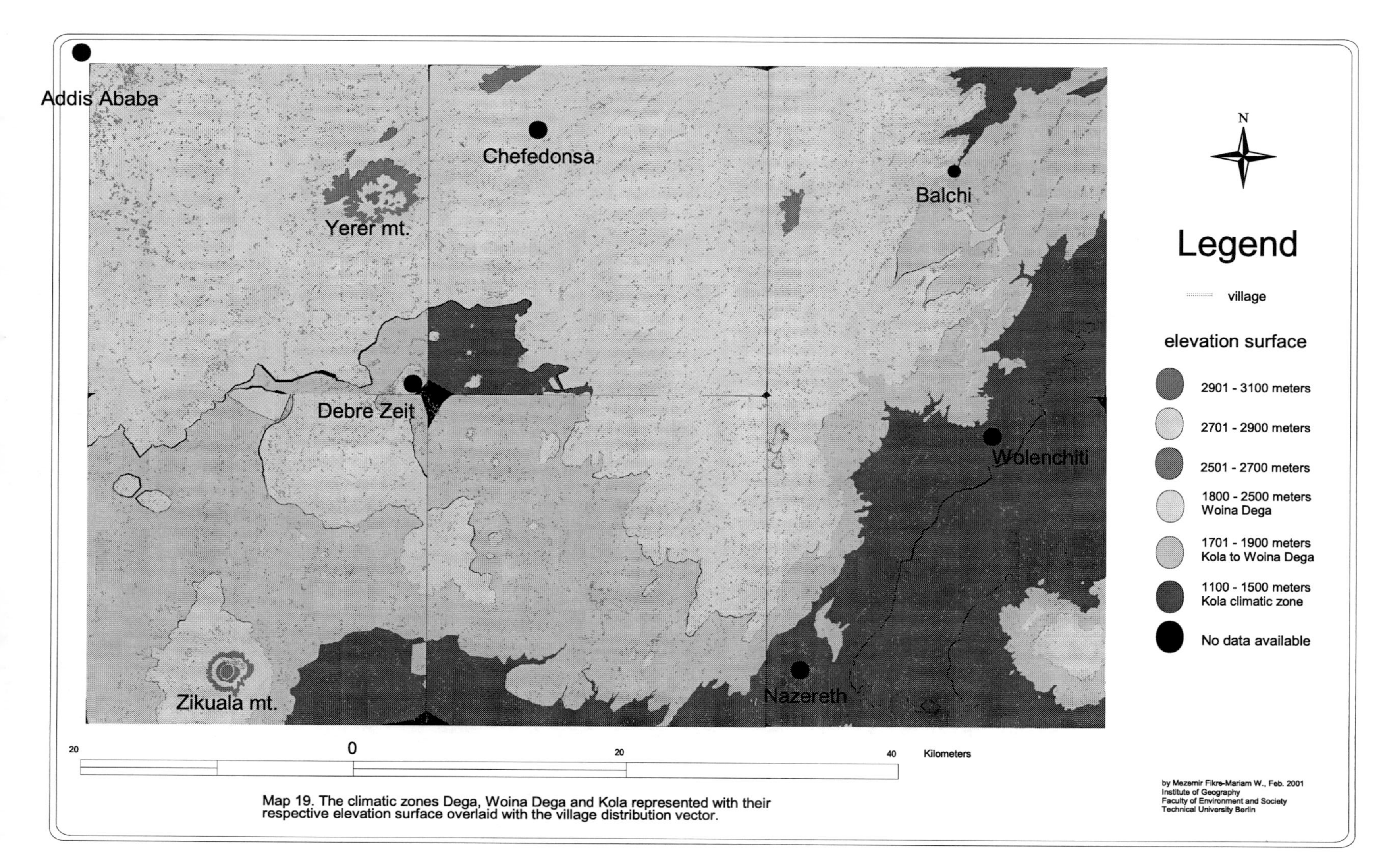

Map 19. The climatic zones Dega, Woina Dega and Kola represented with their respective elevation surface overlaid with the village distribution vector.

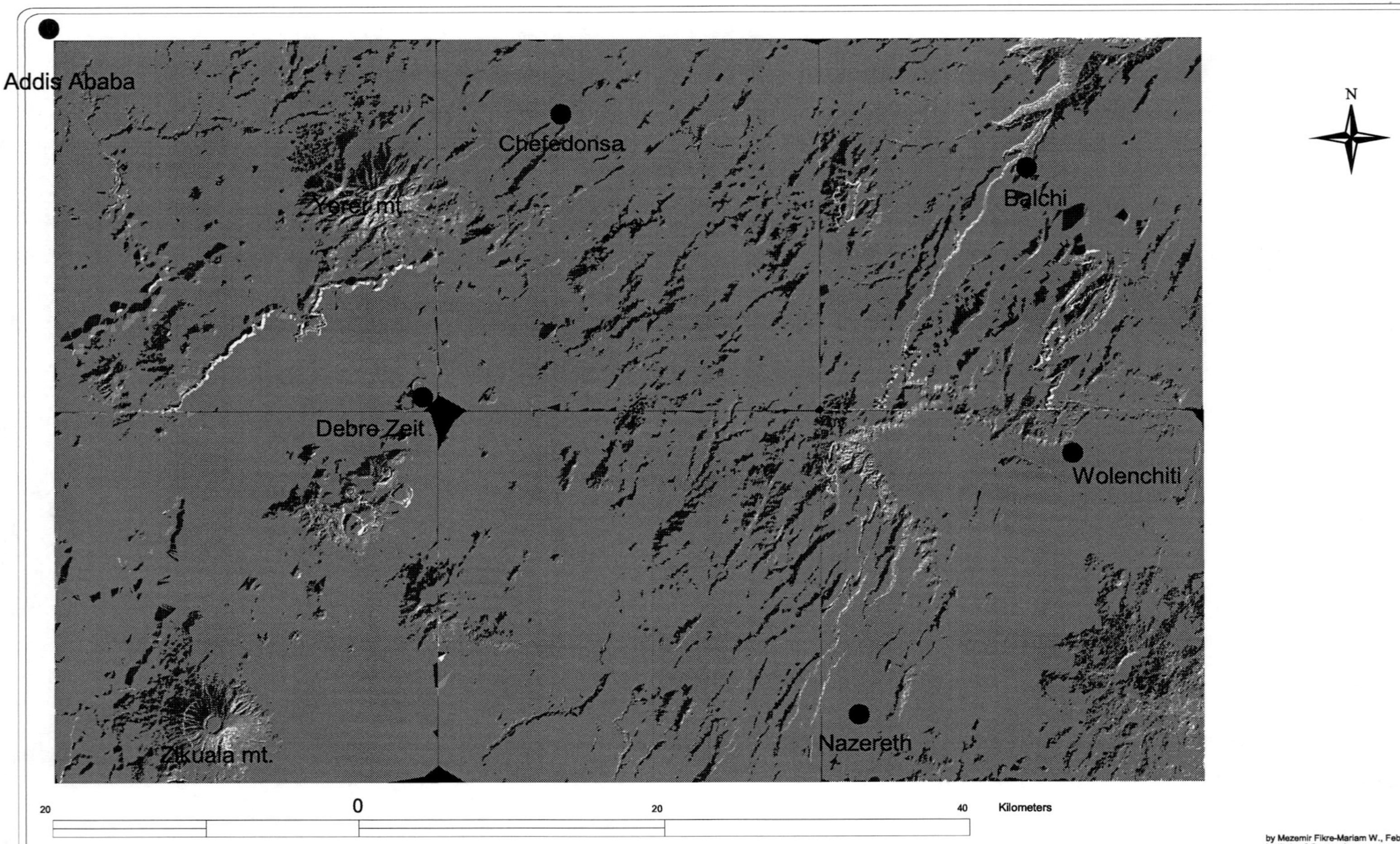

Map 20. Map of shade index with 8 gry-level intervall values.

by Mezemir Fikre-Mariam W., Feb. 2001
Institute of Geography
Faculty of Environment and Society
Technical University Berlin

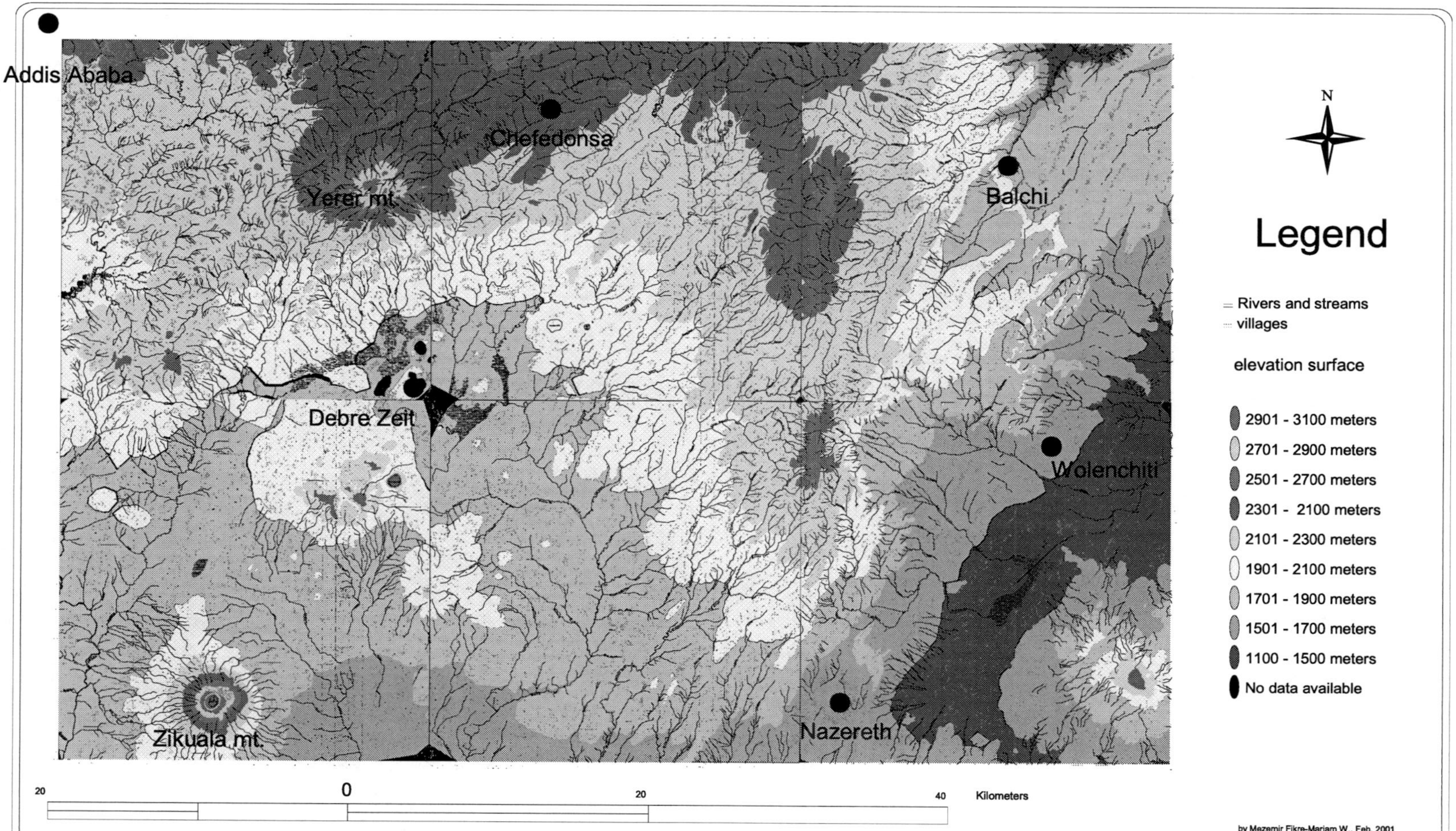

Map 21. Distribution of the villages and the river drainage. The villages are very often far away from main water locations which implies a high water transportation cost.

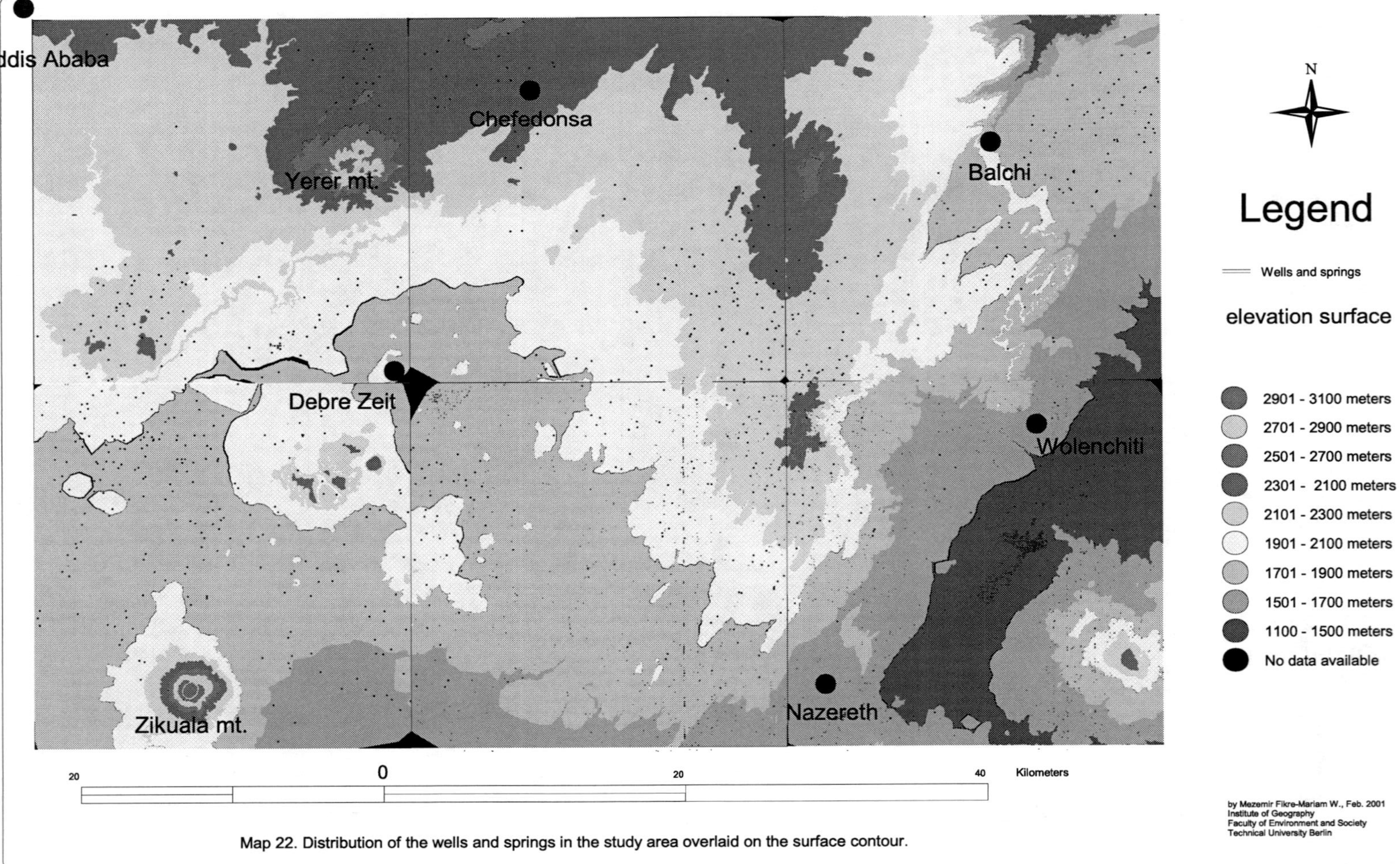

Map 22. Distribution of the wells and springs in the study area overlaid on the surface contour.

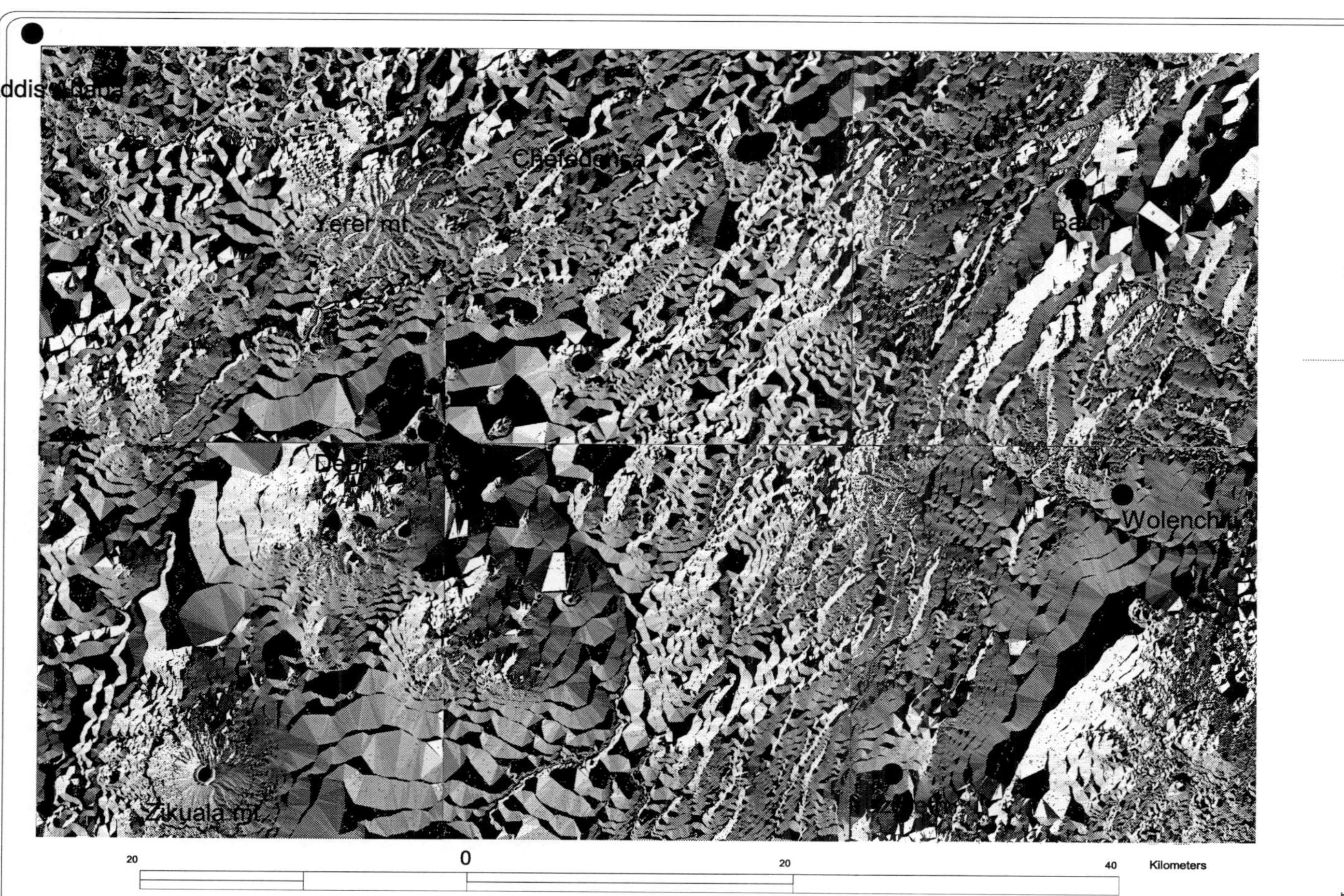

Map 23. North, south, east and west directed aspects overlaid with the village distribution. Often the villages are located on the crossing of the two or more aspects.

by Mezemir Fikre-Mariam W., Feb. 2001
Institute of Geography
Faculty of Environment and Society
Technical University Berlin

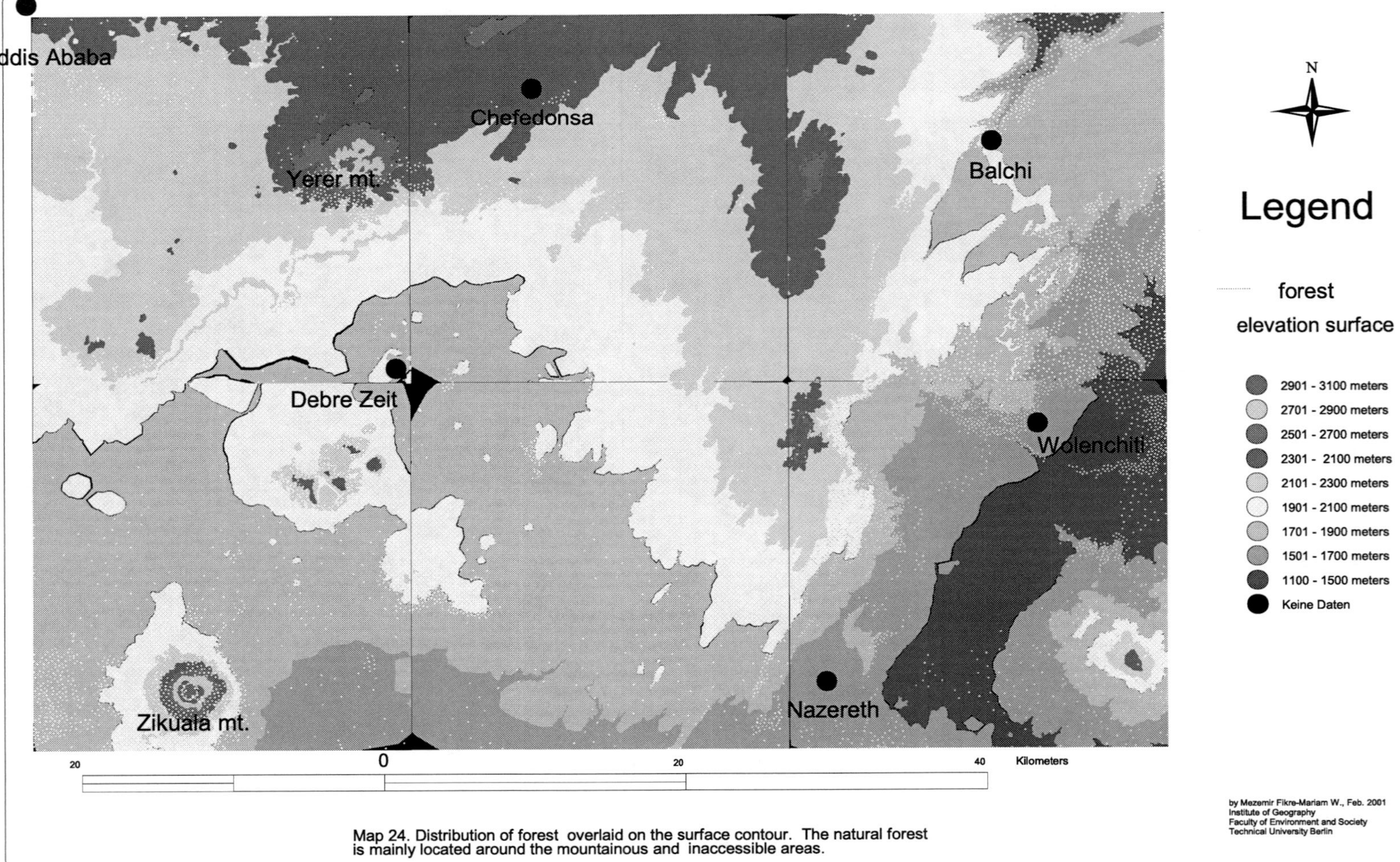

Map 24. Distribution of forest overlaid on the surface contour. The natural forest is mainly located around the mountainous and inaccessible areas.

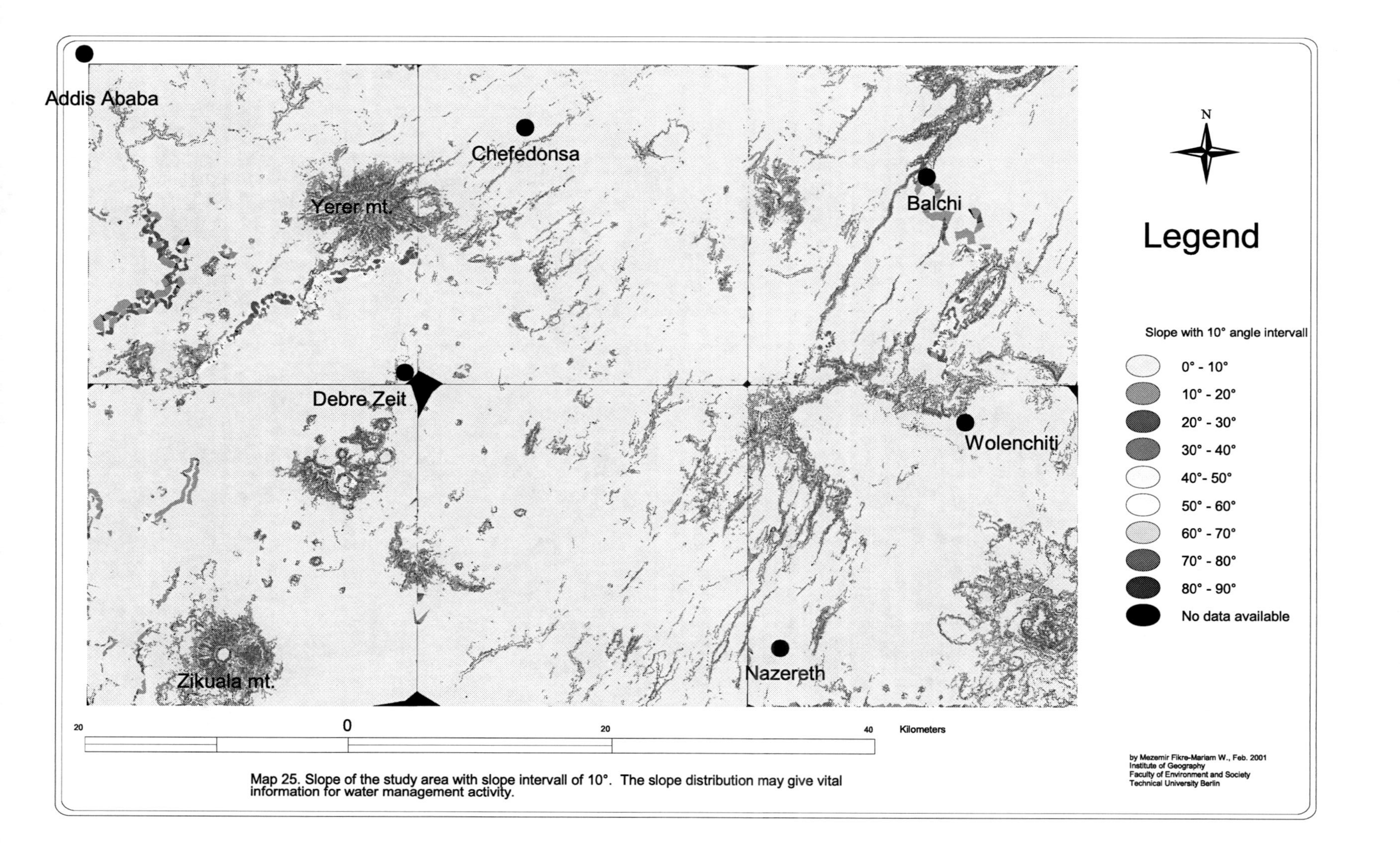

Map 25. Slope of the study area with slope intervall of 10°. The slope distribution may give vital information for water management activity.

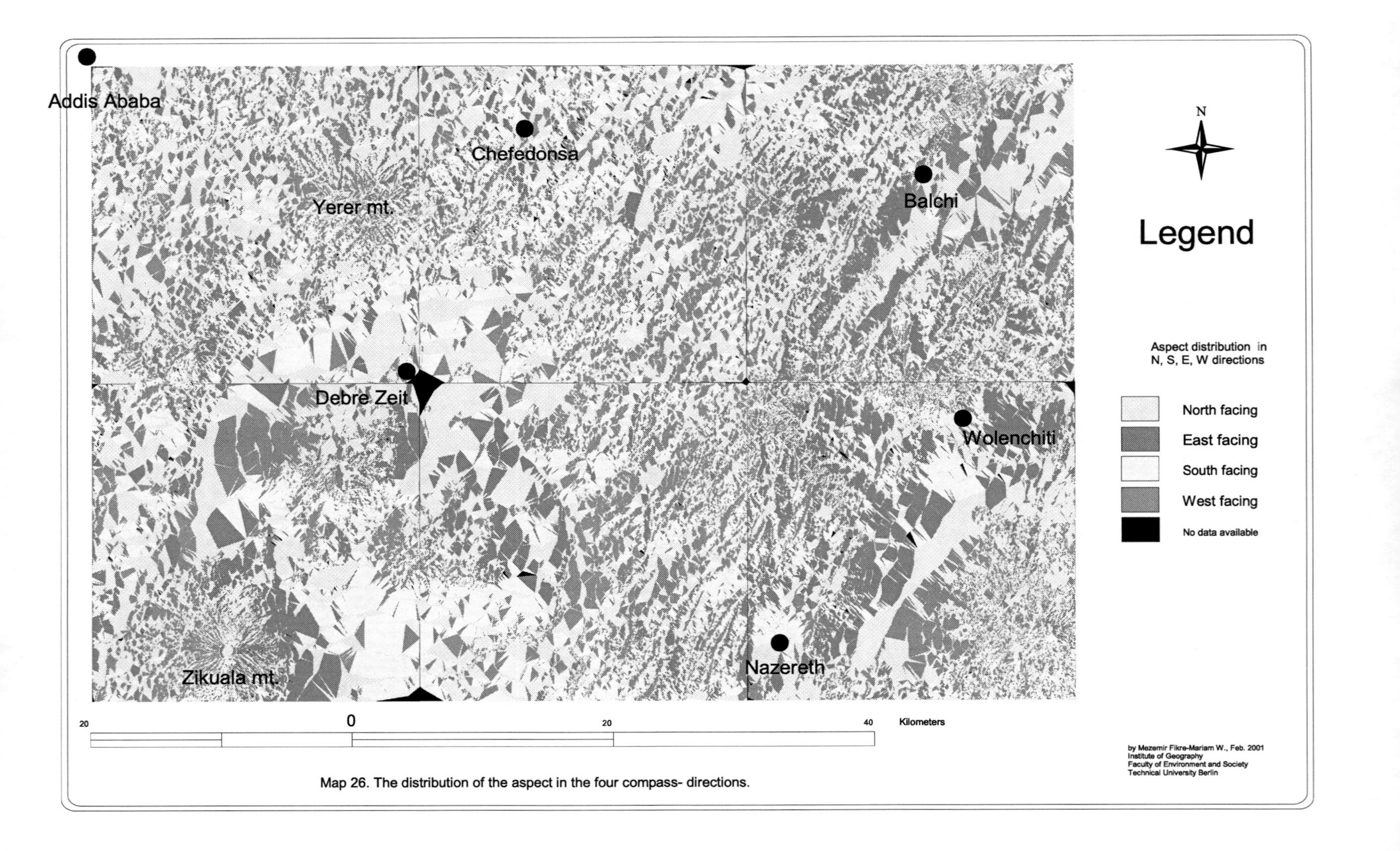

Map 26. The distribution of the aspect in the four compass- directions.

Berliner Geographische Studien

Hrsg.: Prof. Dr. Frithjof Voss, Institut für Geographie der Technischen Universität Berlin
ISSN 0341-8537

40: Exkursionsführer zum 50. Deutschen Geographentag Potsdam 1995. - Hofmeister, Burkhard (Hrsg.); Voss, Frithjof. - 1995. - VI, 423S., 15 Tab., 97 Abb. u. 29 Fotos,17x24 cm. - Br
ISBN **3-7983-1641-4** unverbindl. Preis EUR **2,60**

41: Mäandrierung und Morphodynamik im Eider Ästuar. Am Beispiel der Eider. - Gönnert, Gabriele. - 1995. - XIV, 198 S., 15 Tab., 85 Abb.,17x24 cm. - Br
ISBN **3-7983-1642-2** unverbindl. Preis EUR **2,60**

42: Bürobetriebe und Stadtentwicklung in Berlin. Entwicklungen in Berlin nach 1989 unter besonderer Berücksichtigung der Immobilienbranche. - Acker, Heike. - 1995. - VII, 172 S., 12 Tab., 31 Abb.,17x24 cm. - Br
ISBN **3-7983-1643-0** unverbindl. Preis EUR **2,60**

43: Kommunale Planung im Alto Valle de Rio Negro y Neuquen, Argentinien. Die Rolle der kommunalen Planung bei der Wirtschafts- und Siedlungsentwicklung in der patagonischen Oase. - Albers, Christoph. - 1996. - XII, 244 S., 34 Ktn., 21 Abb., 25 Tab.,17x24 cm. - Br
ISBN **3-7983-1654-6** unverbindl. Preis EUR **12,80**

44: Stadt- und Wirtschaftsraum. Festschrift für Prof. Dr. Burkhard Hofmeister. - Steinecke, Albrecht (Hrsg.). - 1996. - XX, 509 S., 41 Tab., 131 Abb., 16 Fotos, 17x24 cm. - Br
ISBN **3-7983-1686-4** unverbindl. Preis EUR **7,20**

45: Planificación comunal en el Also Valle de Rio Negro y Neuquén, Argentina. - Albers, Christoph. - 1996. - XIII, 243 S., Abb.,Karten, 17x24 cm. - Br
ISBN **3-7983-1696-1** unverbindl. Preis EUR **12,80**

46: Satellite Data Analysis and Surface Modeling for Land Use and Land Cover Classification in Thailand. - Chuchip, Kankhajana. - 1997. - XIV, 238 S., zahl. Abb., Karten, Plaene, 17x24 cm. - Br
ISBN **3-7983-1706-2** unverbindl. Preis EUR **9,70**

47: Integrated Application of the Geographic Information System and Remote Sensing in Solving Hydro-geological and Environmental Problems in the Central Part of Ethiopia and its Possible Extensive Future Use. - Wagaw, Mezemir Fikre-Mariam. - 2002. - 181 S. 17 x 24 cm. - Br
ISBN **3-7983-1739-9** Preis EUR **19,90**

48: Gewässerschutz in Shanghai. - Li, Jian-Xin. - 1997. - IX, 161 S.,21 Abb., 46 Tab., 14 Karten. - Br
ISBN **3-7983-1740-2** unverbindl. Preis EUR **9,20**

49: Nutzungsmischung und Planung. Entwicklung und aktuelle Probleme: Brauereistandorte in Berlin-Prenzlauer Berg. - Oldenburg, Anna. - 1999. - XI, 314 S.,11 Karten, 7 Abb., 17x24 cm. - Br
ISBN **3-7983-1794-1** unverbindl. Preis EUR **21,00**

50: Jungquartäre Geomorphologie und Vergletscherung im östlichen Hindukusch, Chitral, Nordpakistan. - Kamp, Ulrich. - 2000. - XIII, 254 S., davon 18 S. in Farbe, 13 farbige Karten, Tab., Abb., Photos, 17 x 24 cm. - Br
ISBN **3-7983-1812-3** unverbindl. Preis EUR **28,10**

51: Fortführung und Erweiterung von GDF (Geographic Data File) als Datengrundlage für Autonavigationssysteme. - May, Ilka. - 2002. - XXV, 209 S., 17 x 24 cm. - Br
ISBN **3-7983-1897-2** Preis EUR **29,50**

Nicht aufgeführte Bd.-Nrn. sind vergriffen. Bei Abnahme mehrerer Exemplare eines Titels wird Preisnachlaß gewährt; Näheres auf Anfrage. Die Preise sind unverbindlich und gelten für den Barverkauf. Bei Bestellungen wird zusätzlich eine Versandpauschale erhoben: für das 1. Exemplar 2,00 Euro; für jedes weitere Exemplar 0,50 Euro.

Vertrieb/ Publisher: Technische Universität Berlin, Universitätsbibliothek, Abt. Publikationen
Straße des 17. Juni 135, D-10623 Berlin.
Tel.: (030) 314-22976, -23676. Fax.: (030) 314-24741
E-Mail: publikationen@ub.tu-berlin.de

Verkauf/ Book Shop: Gebäude FRA-B - Franklinstr. 15 (Hof), 10587 Berlin-Charlottenburg